WARNINGS

WARNINGS

FINDING CASSANDRAS TO STOP CATASTROPHES

RICHARD A. CLARKE AND R.P. EDDY

ecco

An Imprint of HarperCollinsPublishers

ecco

WARNINGS. Copyright © 2017 by Richard A. Clarke and Randolph P. Eddy. All rights reserved. Printed in the United States of America. No part of this book may be used or reproduced in any manner whatsoever without written permission except in the case of brief quotations embodied in critical articles and reviews. For information, address HarperCollins Publishers, 195 Broadway, New York, NY 10007.

HarperCollins books may be purchased for educational, business, or sales promotional use. For information, please email the Special Markets Department at SPsales@harpercollins.com.

Ecco® and HarperCollins® are trademarks of HarperCollins Publishers.

FIRST EDITION

Designed by Renata De Oliveira

Library of Congress Cataloging-in-Publication Data has been applied for.

ISBN 978-0-06-248802-2

17 18 19 20 21   RS/LSC   10 9 8 7 6 5 4 3 2 1

To Samuel R. Berger,
patriot, counselor, loyal friend, and a man who
listened to Cassandras.
—Richard Clarke

To my wife, Kelly, whom I don't deserve, and to
our children Reed, Tucker, and William, who
don't deserve a broken world.
Let's fix it.
—R.P. Eddy

CONTENTS

WARNINGS

CHAPTER 1

Cassandra: From Myth to Reality

There are people among us who can see the future.

Often they clamor for our attention, and just as often they are ignored. We are right to discount most soothsayers, but horrible things happen when accurate warnings of specific disasters go unheeded. People die because we fail to distinguish the prophet from the charlatan.

This book tries to find those rare people who see the future, who have accurate visions of looming disasters.

Cassandra was a beautiful princess of Troy, cursed by the god Apollo. He gave her the ability to see impending doom, but the inability to persuade anyone to believe her. Her ability to pierce the barriers of space and time to see the future showed her the fiery fall of her beloved city, but the people of Troy ridiculed and disregarded her. She descended into madness and ultimately became one of the victims of the tragedy she foretold.

Are there Cassandras among us today, warning of ticking disasters, whose predictions fall on deaf ears? Is it possible to figure out who these seers are? Can we cut through the false warnings to tune in to the correct visions, saving millions of lives and billions of dollars? That question is not about Greek mythology. It is about our ability today, as a nation, as an international community, to detect impending disaster and act in time to avoid it, or at least to mitigate the damage.

Buried in billions of pages of blog posts and tweets, academic research, and government reports, Cassandra figuratively calls to us, warning of calamity. Often she is unheeded, sometimes unheard. Frequently she is given only a token response or dismissed as a fool or a fraud. Her stories are so improbable, so unprecedented, that we cannot process them or believe them, much less act upon them.

The problem is, of course, that Cassandra was right, and those who ignored her may have done so at the cost of their own lives and that of their state.

Not just the ancient Greeks had a tale of a seer tragically ignored. The Bible tells the story of the Hebrew prophet Daniel, who was able to read mysterious words that appeared on the wall of the Babylonian king Belshazzar's banquet hall during a rowdy feast. The words *mene, mene, tekel, upharsin* ("numbered, numbered, weighed, divided" in Aramaic) were unintelligible to all but Daniel, who warned the king that they foretold the fall of his kingdom. According to the story, Belshazzar was killed in a coup only hours later. Daniel had seen the "writing on the wall."

Today when someone is labeled a Cassandra, it's commonly understood that they simply worry too much and are fatalistic, overly pessimistic, or focus too much on the improbable downside, a Chicken Little rather than a prophet. If we refer to the original Greek myth, a Cassandra should be someone whom we value, whose warnings we accept and act upon. We seldom do, however. We rarely believe those whose predictions differ from the usual, who see things that have never been, whose vision of the future differs from our own, whose prescription would force us to act now, perhaps changing the things we do in drastic and costly ways.

What the ancient Greeks called Cassandra behavior today's social scientists sometimes refer to as sentinel intelligence or sentinel behavior, the ability to detect danger from warning signs before others see it. The behavior is observed in a variety of animals, including, we believe, in humans. Those with sentinel intelligence see with great clarity through the fog of indicators, and they warn the pack. In other animals, the pack seems genetically disposed to respond quickly to the warnings of their sentinels. In humans, that ability is less well developed.

We, the authors, are Dick Clarke and R.P. Eddy. We have known and worked with each other for over twenty years, in and out of government, on topics including the rise of al Qaeda and then ISIS, nuclear and biological weapons proliferation and the emergence of deadly viruses like Ebola and HIV, the introduction of the cyber threat, and wars from Iraq to Bosnia to Afghanistan. We are neither pessimists nor obsessed with doom. Indeed, we are optimists who believe we are on the cusp of great technological advances that should make human life vastly better. For much of the time we have known each other, however, the news has been dominated by a series of disasters that could have been avoided or mitigated. Among them are 9/11, the Great Recession, Katrina, the Second Iraq War, and the rise of ISIS. We worked directly on many of these topics, and were personally affected by some. They have loomed large in our conversations with each other, often over a good single malt, with one question regularly recurring: how could we have avoided or better prepared for that?

What we noticed was that almost all of these events were followed by investigations and recriminations, seeking to lay blame for the catastrophe or responsibility for failures that made the situations worse.

In many instances, however, it seemed that an expert or expert group, a Cassandra, had accurately predicted what would happen. They were often ignored, their warnings denigrated, disregarded, or given only inadequate, token responses.

We began to wonder whether there was some pattern of prescient but overlooked warnings. Could there possibly be a way to identify these accurate but unheeded warnings before disaster struck? If there really was a frequent phenomenon of unheeded alarms that later proved to be accurate, finding a way to detect and validate those warnings in advance could save lives, avoid suffering, and reduce financial losses.

We discovered that at any given time there is a plethora of predictions of doom. Most are ignored because they should be. They are created by cranks and have no empirical underpinnings or basis in reality. Some warnings are heeded, but then events prove the alarm to be false. Often, however, true experts in a field do their job and sound the warning in time, only to be ignored or given only an inadequate, token response. We began calling such episodes Cassandra Events. That led us to ask whether there was something about past Cassandra Events that can help us identify contemporary alarmists whose warnings will turn out to be right. We asked friends, colleagues, associates, and world-leading experts what they thought about this Cassandra phenomenon.

The Cassandra problem is not only one of hearing the likely accurate predictions through the noise, but of processing them properly once they are identified. We began to realize that to successfully navigate a Cassandra Event, an organization or society must move through several stages. First we must hear the forecast, then believe it, and finally act upon it.[1] In practice, these steps are each individually challenging. Moreover, executing all three sequentially is often immensely difficult. In particular, the ability to get it right is exceedingly rare when the prediction varies substantially from the norm, from the past, from our experience, or from our deeply held beliefs about the way the future *should* unfold. Add a significant financial cost as a requirement of acting on such a warning, and the probability for action often approaches zero. If, however, we ignore a true Cassandra, the cost of not acting is usually far higher than the cost of dealing with the problem earlier.

Thus, this book will seek to answer these questions: How can we detect a real Cassandra among the myriad of pundits? What methods, if any, can be employed to better identify and listen to these prophetic warnings? Is there perhaps a way to distill the direst predictions from the surrounding noise and focus our attention on them? Or will Cassandra forever be condemned to weep as she watches her beloved city of Troy burn?

To answer those questions, we begin with short case studies of real human beings in current times who had a Cassandra-like ability concerning some important issue, and who, like the mythological princess, were ignored. This book will not attempt to be the definitive case study of any of the disasters we review. Instead, we will focus on the Cassandras themselves and their stories. We will try to determine how they knew when others did not, why they were dismissed, and how circumstances could have been changed so that their warnings would have been heeded.

While our case studies focus on individuals, many are also stories about organizations created to be sentinels. Governments and some industries and professions have long realized the value of having lookouts, scouts, and sentinels to give warnings. Among our case studies are stories involving parts of the U.S. government: the intelligence community's "warning staff," the agency created to watch for potential mine disasters, the national disaster management organization, and the financial regulators who exist to look for fraud and potential systemic economic instability.

We begin our examination of past Cassandra Events by looking at Saddam Hussein's invasion of Kuwait in 1990, an event that gave rise to a series of disasters that continue today. Next, we examine what happened in Louisiana before Hurricane Katrina. We examine the concurrent calamities of a tsunami and multiple nuclear reactor meltdowns in Japan. Back in the United States, we go underground in West Virginia to take a look at the recurrent nature of mining disasters. The Middle East intrudes again with the rise of Daesh (ISIS). Then we shift to economic and financial Cassandra Events, examining the case of Bernie Madoff's Ponzi scheme, as well as the Great Recession of 2008.

As we proceeded through these Cassandra Event case studies in a variety of different fields, we began to notice common threads: characteristics of the Cassandras, of their audiences, and of the issues that, when applied to a modern controversial prediction of disaster, might suggest that we are seeing someone warning of a future Cassandra Event. By identifying those common elements and synthesizing them into a methodology, we create what we call our Cassandra Coefficient, a score that suggests to us the likelihood that an individual is indeed a Cassandra whose warning is likely accurate, but is at risk of being ignored.

Having established this process for developing a Cassandra Coefficient based on past Cassandra Events, we next listen for today's Cassandras. Who now among us may be accurately warning us of something we are ignoring, perhaps at our own peril? We look at contemporary individuals and their predictions, and examine the ongoing public reaction to them. Our cases here include artificial intelligence, genetic engineering, sea level rise, pandemic disease, a new risk of nuclear winter, the Internet of Things, and asteroid impacts.

Finally, we end this volume with some thoughts about how society and government might reduce the frequency of ignoring Cassandras when it comes to some of the major issues of our time.

While we will not endorse the predictions of the possible contemporary Cassandras (we leave it to the reader to decide), we will apply our framework to their cases, evaluating each element—the individual, the receiver of the warning, and the threat itself—to determine the Cassandra Coefficient. If you find value in our methodology, perhaps we would do well as a society to turn our attention to those scenarios with the highest scores: the future high-impact events we are ignoring.

If this seems like an ambitious undertaking, we don't think so. It's actually quite easy to put the pieces together. The hard part may be to first realize these invisible people exist.

Until *Warnings*, no author has explored how to sift through the noise to identify the actual Cassandras. This is the first effort to help people judge which warnings deserve a closer listen, and thereby perhaps stop these disasters before they happen.

A FEW PEOPLE THOUGHT WE WERE INTERESTED IN THIS TOPIC BE-cause of Dick's role in the September 11 tragedy. "So, are you interested in Cassandras because you, Dick Clarke, were the Cassandra about al Qaeda and 9/11?" The short answer to that question is no. Both authors have an interest in the phenomenon of Cassandras because of our fascination with leadership decision making and its role in significant historical events and trends. Inevitably, however, people want to talk about al Qaeda and the attacks of September 11, 2001. Dick addressed this issue in detail many years ago in his book *Against All Enemies*, and neither of us has any desire to replow much of that ground again here, so let's just get it out of the way. Here is what Dick thinks about warnings and 9/11:

A lot of other people were also warning about the al Qaeda threat by 1999 and certainly by 2001. Chief among them were FBI Special Agent John P. O'Neill, CIA Director George Tenet, and most of the leadership of the CIA's Counter-Terrorism Center. Regrettably none of us was able to predict the time, place, or method of the September 11 attacks. We had achieved what military intelligence people call strategic warning—identifying intent—but not tactical warning—identifying the when, where, and how. Why we failed to achieve tactical warning is a controversial subject centering on the repeated, conscious decision of senior CIA personnel to prevent the FBI and White House leadership from knowing that the 9/11 hijackers were already in the U.S. (A much more detailed discussion of the topic can be found in another of Dick's previously published books, *Your Government Failed You.*)

The Bush administration's response to the warnings leading up to 9/11 mirrors the bungled responses we will discuss further in the Cassandra Event case studies presented in this book. Officials heard the warnings but didn't fully believe them and certainly didn't act on them. Most of the leadership had been out of government since the previous Bush administration eight years prior. Their biases were a decade old. They couldn't believe that the world had changed so much. The number-one threat to the United States was not a nation-state but a stateless terrorist group? Much less could they believe that the threat would manifest as a plan to attack the country from within. Simply put, it had never happened before, so they couldn't really believe that it would.

Unfortunately, even when recognized experts and institutions explicitly and loudly sound the alarm, they do not always succeed in effectively conveying a message or eliciting a meaningful response from the appropriate authorities. Decision makers don't typically welcome predictions of impending disaster. Rather than acting as comprehensively as possible to prevent or mitigate the effects of the coming catastrophe, they often go into an implicit state of denial. They may not dispute the evidence and reject the warning, but they don't act as though they actually believe it to be true. So it was with the al Qaeda threat, the terrorist attacks of September 11, 2001, and the Bush administration. Now let's move on.

IN OUR PRELIMINARY DISCUSSIONS WITH FRIENDS AND COLLEAGUES, we were directed to a couple of famous examples of Cassandras in recent history. The most common figure people repeatedly brought up was the historic giant Sir Winston S. Churchill. Thick volumes on Churchill's public and private life abound, some written by the man himself. Indeed, the former British prime minister was a prolific historian. Few public figures have documented and explained their lives and careers at greater length than Churchill. It is likely that no others have been the subject of so much scholarly research and writing, excepting perhaps U.S. President Abraham Lincoln. Consequently, we know a lot about Churchill's life, including those events that might qualify him as a Cassandra. We also have a rich body of work describing the man's disposition, intellect, and character.

Unlike some Cassandras whom we will meet later in this book, Churchill was far from an obscure, unknown individual when he began his warnings in the 1930s. He had been elected to Parliament in 1900 at age twenty-six and became a Cabinet member a mere eight years later. He served in a succession of ministries in domestic policy, economic affairs, and foreign policy. By the middle of the next decade, he was best known for his role as the civilian in charge of the Royal Navy during World War I. He was the mastermind behind an attack on Turkey that was poorly implemented, resulting in the British defeat at Gallipoli. Resigning under criticism, he joined the army and went to the Western Front, the trenches in France.

He later attracted more attention by switching political parties, making him an even more controversial figure. In the 1930s, Prime Minister Stanley Baldwin and his successor, Neville Chamberlin, both thought Churchill a troublemaker. His writings, books, and newspaper columns focused on military topics, contributing to his image as a militarist. Thus, when in 1933 he began warning of the growth of German military power and calling for expansion of the British military, many discounted what he had to say.

Unlike many Cassandras who themselves generated the empirical data that drove them, Churchill did not discover the extent of the German rearmament problem on his own. Concerned civil servants and military officers fed him secret government documents. One source in particular provided Churchill details on the rapidly expanding gap between the Royal Air Force and the new Luftwaffe. Another source passed him photographs of a new wall of German fortifications and defenses, the Siegfried Line, which would make it difficult, if not impossible, for France and Britain to invade or counterattack. Armed with this smuggled intelligence, Churchill railed against the German air threat in speeches to Parliament, predicting that in a future war the Germans would sweep through Belgium and Holland into France. Many government ministers dismissed his claims and the data, despite the intelligence having come from within the government itself.

Some in government did take Churchill's warnings seriously even before they were proven true. In the mid-1930s, a friendly member of Parliament and a future prime minister, Harold Macmillan, concerned that Churchill would give up, cautioned that he was "in danger of relapsing into a complacent Cassandra." Macmillan described Churchill's attitude at the time as, "I have done my best. I have made all these speeches. Nobody has paid any attention. All my prophecies have turned out to be true. I have been publicly snubbed by the

Government. What more can I do?"[2] Lady Longford, the famed British historian, even thought at the time that Churchill was "the disregarded voice of Cassandra."

What Churchill did have in common with many others who accurately warned of impending disasters was a colorful personality, often described as compulsive, driven, hardworking, outspoken, and abrasive. He was thought of as a permanent hawk and an adventurer. Macmillan noted that there was "general doubt as to the soundness of his judgment." The person who ultimately saved Churchill's reputation was Hitler, whose actions gradually proved the outspoken British critic correct. As Hitler's aggression grew, Churchill was asked to rejoin the British Cabinet. When war clouds on the horizon grew plain to see, he was asked to become prime minister.

As a wartime leader, Churchill was masterful, keeping the spirits of the beleaguered people from flagging as the nation's fortunes slipped, and bringing creativity to military strategy, tactics, and technology. Some historians consider his contributions the key factor between victory and defeat. Yet his prescience and leadership were not enough to carry him past the end of World War II. Churchill was defeated in 1945, in the first election after the war.

A more contemporary Cassandra was the engineer who fought the NASA leadership prior to the 1986 explosion of the space shuttle *Challenger*. Indeed, the Cassandra Event involving Roger Boisjoly and his attempts to prevent the shuttle launch on that fateful morning in January 1986 has become one of the preeminent case studies in risk management and decision-making ethics.

The *Challenger* disaster stemmed from an inherent flaw in the original design of the solid rocket boosters used to launch the Space Shuttle into orbit. Morton-Thiokol, the company awarded the contract to build the boosters by NASA, had modeled its design on the reliable Titan III rocket. The cylindrical booster sections were manufactured separately, then mounted end-on-end, using O-rings and putty to seal the sections together. However, Morton-Thiokol made several significant changes to the Titan III design to simplify the complicated manufacturing process and cut costs, including changing both the way the sections were mated and the orientation of the O-ring seals.

During initial testing and later, even after the Shuttle started flying, engineers at Morton-Thiokol and NASA became increasingly alarmed by problems discovered in the modified design of the rocket booster. Rather than creating an airtight seal, hot combustion gases were burning through the putty, leaking into the joints, and burning the O-rings. Moreover, this problem, which became known as joint rotation, was exacerbated by cold temperatures.

Boisjoly was a solid rocket booster engineer at Morton-Thiokol. He became seriously concerned, not only by the joint rotation itself, but by the fact that NASA continued pressing ahead with Shuttle launches with the consent of the Morton-Thiokol management, despite not having found an acceptable fix. Among top officials in both organizations, the O-ring problem became viewed as an acceptable flight risk. In a memo dated July 31, 1985, about six months before the *Challenger* incident, Boisjoly warned Thiokol's vice president of Engineering that the O-ring issues were a disaster in the making, ominously predicting that the "result would be catastrophic of the highest order—loss of human life." Boisjoly's warnings were ignored.

The morning before *Challenger*'s final flight, the temperature forecast was 30 degrees Fahrenheit, far colder than any previous Shuttle launch. Boisjoly and his engineering colleagues at Thiokol argued that, due to past O-ring performance problems, the temperature posed an unacceptable risk to the safety of the launch. These Cassandras were unable to persuade NASA and Morton-Thiokol's management, who were under pressure after several previous aborted launch attempts.

The launch proceeded on the morning of January 28, 1986, in an ambient temperature of 36 degrees Fahrenheit, 15 degrees colder than on any previous attempt. Challenger's mission ended at 11:39 a.m., seventy-three seconds after liftoff, when an O-ring seal in the right solid rocket booster failed. The result was complete structural failure. A horrified American public, along with engineers and managers at NASA and Morton-Thiokol, watched as the vehicle was ripped apart in a giant fireball.

As we will see, Cassandras are seldom appreciated for their efforts even after the disaster comes to pass. Roger Boisjoly was shunned and ostracized by colleagues and managers at Morton-Thiokol, who thought that his testimony during the accident investigation would cost them their jobs. We can, perhaps, cynically understand such a reaction as human nature, an instinct for survival. However, such a response does little to help prepare for future disaster or to ensure that Cassandra's warnings do not again go unheeded.

As it seemed we weren't the only ones who had noticed Cassandras in our midst, we then wondered if there existed any scholarly research on the topic of predictions. In fact, prediction is something that academics have spent a lot of time studying and considering. The statistician Nate Silver has taken a highly quantitative approach to prediction, one that works for a certain class of event. The jurist Richard Posner examined the phenomenon of catastrophes in the years after 9/11. Psychologists like Dan Ariely and Tsachi Ein-Dor have probed the way our brains work (and don't) through empirical observation and the study of warnings.

Unquestionably one of the foundational works in this area, predictions within the social sciences, is Philip Tetlock's *Expert Political Judgment*. Tetlock began his project in the 1980s, somewhat tentatively at first, and then with increasing rigor. Then a psychology professor at UC Berkeley, he felt that the American public was awash in predictions by "experts," especially on issues like national security and economic policy. He endeavored to make sense of the many predictions, which he saw were all too often presented by self-assured experts as inevitable, only to directly contradict a prediction made by yet another self-assured expert.

Over the next nearly twenty years, Tetlock asked a panel of 284 experts, defined as "a professional who makes his or her livelihood by commenting or offering advice on political and economic trends of significance," to make predictions on a variety of issues, both within and outside of their specific areas of expertise. While the study was carried out anonymously, the experts were generally all highly educated with at least several years of experience in their field. Some worked in government, some in the private sector, international agencies, or think tanks. Tetlock compiled and analyzed their responses, over twenty thousand in all, and then made some fascinating predictions of his own.

Overall, experts were terrible at forecasting the future, but Tetlock did something interesting: in addition to asking the experts *what* they thought about a particular scenario, he also examined *how* they thought. After evaluating their cognitive styles, Tetlock divided the experts into two categories, "hedgehogs" and "foxes," after an essay written by the philosopher Isaiah Berlin. In his essay, Berlin references the Greek poet Archilochus: "The fox knows many things, but the hedgehog knows one big thing." The hedgehog style of thinking is marked by a belief in the truth of "one big thing"—such as a fundamental, unifying theory—and then "aggressively extend[ing] the explanatory reach of that one big thing into new domains." Conversely, a fox's cognitive style is marked by "flexible 'ad hocery' that require[s] stitching together diverse sources of information." Tetlock writes that foxes are "rather dubious that the cloudlike subject of politics can be the object of a clocklike science."

Once the experts were separated in this way, Tetlock discovered some surprising relationships. The foxes consistently beat the hedgehogs in the accuracy of their predictions by a significant margin. Not only did the foxes perform better when asked forecasting questions within their areas of expertise, they still performed better than hedgehogs on questions with which they were less familiar. And perhaps most surprisingly of all, hedgehogs actually did significantly worse when asked to make predictions *within* their areas of expertise. Tetlock attributed this startling finding to the idea that "hedgehogs dig themselves into intellectual holes. The deeper they dig, the harder it gets to climb out and see what is happening outside, and the more tempting it becomes to keep on doing what they know how to do . . . uncovering new reasons why their initial inclination, usually too optimistic or pessimistic, was right."

Still, maddeningly, even the foxes, considered as a group, were only ever able to approximate the accuracy of simple statistical models that extrapolated trends. They did perform somewhat better than undergraduates subjected to the same exercises, and they outperformed the proverbial "chimp with a dart board," but they didn't come close to the predictive accuracy of formal statistical models.

Later books have looked at Tetlock's foundational results in some additional detail. Dan Gardner's 2012 *Future Babble* draws on recent research in psychology, neuroscience, and behavioral economics to detail the biases and other cognitive processes that skew our judgment when we try to make predictions about the future. And building on a successful career in sports and political forecasting, Nate Silver discusses in his book, *The Signal and the Noise,* how thinking more probabilistically can help us distill more accurate predictions from a sea of raw data.

Fundamentally, these books all identify the difficulties inherent in trying to see into the future. Predicting natural phenomena is stymied by the chaotic nature of the universe: natural processes are nonlinear systems driven by feedback loops that are often inherently unpredictable themselves. Human behavior, or the behavior of a government or society, is driven by countless factors, making behavioral forecasting all the more difficult.

Thus, our initial and informal survey of those whose judgment we trust told us that people had thought about related issues and they even had in mind some archetypal Cassandras. Some experts had published learned books that plowed the ground on the phenomenon of prediction in science, social science, engineering, and secret intelligence. None of the books we researched really focused on the experts who saw disaster coming or why they were ignored. And none asked how we can spot those with sentinel intelligence before a disaster occurs. Therefore we do so in this book.

One reason that no one has focused on the people who warn is that many, probably most, prophets are wrong, giving the others a bad reputation. We have listened to forecasts of the world running out of fossil fuels, being unable to grow enough food for humanity, being crushed by a wildly proliferating population. Today, we look around and see a glut of hydrocarbons, new ways to increase crop yields, and some nations now worrying about negative population growth. Whom to believe?

Another reason that people do not pay attention to these kinds of prophecies, we postulate, is that even putting aside the numerous erroneous warnings, there is just a lot of data, an ever expanding sea of it. People and organizations often try their best to make sense of it to inform their decision making, perhaps through "big data analytics" and machine learning. It's an important effort for many businesses and governments: knowing which seemingly low-probability prediction accurately forecasts a high-impact event can be the difference between success and failure, sometimes between life and death.

There is also an overwhelming amount of seemingly well-informed expert opinion coming at us from endless channels on the Internet, television, and print. Blinded by the bright promise of big data and awash in the cacophony of pundit opinion, how can a decision maker identify Cassandra's warning?

For us, the authors, this has been a journey of discovery, exploring a wide variety of important issues that we found fascinating and gripping. It has meant, in many cases, sitting down with the candidate Cassandras and getting to know them as people, as well as discussing their issues. It has taken us on physical journeys from the shores of the Arab Gulf (or Persian Gulf, if you prefer) to laboratories in California. It has taken us on mental journeys from rocks spinning beyond the Solar System to rocks deep underneath the mountains of Appalachia. We can already tell you this much: if even just *some* of these contemporary warnings are right, we are in for a rough future. Thus, it is important for all of us to decide whether or not we should be doing more now to mitigate these potential crises.

If we are right, searching for Cassandras using this systematic methodology could help today's leaders reprioritize their attention and resources in a meaningful way to address the threats of tomorrow. *Warnings* might allow people who would have otherwise been Cassandras to become the heroes who convinced society to act in time.

We begin this journey well aware that, like the citizens of ancient Troy, we are mere mortals frequenting the precincts of the oracular temples of the gods in our search for the truth and a vision of the future. We know that inevitably we will err. Nonetheless, we try and try again, lured toward the prospect of foreseeing and preventing disasters on the horizon. We seek to understand: to learn how to listen for Cassandra.

CHAPTER 2

The Spook:
Invasion of Kuwait

In conditions of great uncertainty people tend to predict the events that they want to happen actually will happen.
—ROBERTA WOHLSTETTER, *PEARL HARBOR: WARNING AND DECISION*

Charlie Allen wasn't planning a summer vacation. He hadn't taken one in years. July was just like any other month to him, days that began with him arriving at the office around five thirty a.m. and leaving some fourteen hours later. There were so many intelligence reports to read, so much intelligence collection to task. He didn't want to miss anything, and he certainly did not want to be surprised. Some colleagues thought him compulsive, obsessive, but he had not made it this far by overlooking any information of significance. Weeks, months, and years of work had culminated in this moment, a moment when he would be forced to exercise his authority to the fullest extent. It was a risky move, but he had been selected for this position precisely because he had such resolve. He let the intelligence guide his actions, and the intelligence had clearly spoken. Charlie reached for a red pen and his legal pad and, in big block letters, began to scrawl out the words, "Warning of War" at the top. This message would be delivered directly to the President of the United States.

That July, in 1990, Charlie was fifty-four and had worked exactly half of his life for the Central Intelligence Agency. Technically, he had been on loan from the CIA since 1986 to something called the National Intelligence Council (NIC). The NIC was a small organization housed in CIA headquarters, but in the larger government hierarchy, it actually sat above the dozen or more intelligence agencies. The Council was comprised of ten national intelligence officers (NIOs), each of whose job it was to coordinate among the various agencies on one set of issues and report anything significant to the White House. But Charlie's issue was not a single issue at all. Charlie was the man in charge of an ominous portfolio called "Warning."

That morning, July 25, Charlie had been at work for over an hour, but the CIA parking lot was still empty, as were most of the offices in the headquarters complex. He was on his second cup of coffee, staring at a stack of photographs of Iraqi Republican Guard divisions that had initially moved south toward the Kuwaiti border on July 15. The camera that had taken their pictures was hundreds of miles up in space, spinning around the globe at thousands of miles an hour, and Charlie Allen had aimed it. Before he had left the office the night before, Charlie had ordered the satellite to image those elite armor divisions. Now, looking at the results of the imagery collection, he did not like what he saw.

As the National Intelligence Officer for warning, Charlie's job was the institutional embodiment of national security lessons painfully learned. His position existed so that the nation would not be blindsided again as it been by past crises, most notably at Pearl Harbor. There were five separate government investigations of why the United

States had not seen the Japanese attack coming. A joint Senate-House probe resulted in a thirty-nine-volume report.

It was not until 1962 that the definitive report on Pearl Harbor was written. And then, it was not written by a government committee, but by an academic named Roberta Wohlstetter. She concluded that the problem had not been a scarcity of information, but an overabundance of it. It was difficult, Wohlstetter concluded, to identify the reports that mattered among the avalanche of intelligence and other inputs. Analysts in many military organizations had been inured to the steady drum roll of new intelligence coming in, but no one had the job of putting it all together or issuing a warning to alert the rest of the government if and when an overarching tipping point had been reached. There had been no national intelligence officer to interpret as a whole all of the intelligence the United States possessed. There had been no national intelligence officer for warning who could decide on his own to hit the alarm button before it was too late.

Dick Clarke first met Charlie Allen during the Reagan administration. Every Thursday morning in a nondescript building near the White House, Charlie would assemble a group of senior analysts from the major U.S. intelligence agencies. Allen dubbed the group the Warning Committee. Clarke represented the State Department's intelligence unit, an outgrowth of the analytical part of the legendary World War II spy outfit known as the Office of Strategic Services, the OSS.

Clarke would walk alone to the unlabeled building carrying a locked briefcase that contained the intelligence reports that had bothered him most during the week. (He thought it was probably a security violation to walk down the street with those intelligence gems, so he never asked permission.) Every member of Charlie Allen's committee brought their own concerns to the Warning Committee, and the result was something resembling a group therapy session. By the end of the day on Thursday, Allen produced a punchy two-page memorandum informed by the Warning Committee meeting. It was a list of things to watch, trends, military exercises, potential troublemakers. The meetings also resulted in "collection taskers," orders for some satellites to take pictures and others to listen to certain frequencies, requests for certain coded messages to get priority decryption, and suggestions that spies focus on particular targets.

The challenge the Warning Committee faced, Allen would remind them frequently, was to not be "the boy who cried wolf." They should be measured in their assessments and the language they used. It was also imperative that the weekly report be something that intelligence managers wanted to read, something that helped senior people focus, not a scolding that they had missed something important. In theory, the Weekly Warning Report was distributed to the President and all senior intelligence managers, but no one actually deluded himself into thinking that President Ronald Reagan was an avid reader of the committee's work.

Charlie Allen did, however, have a way of getting directly to the President. As the NIO for warning, he had a special authority, one that he had never before used. If, in his assessment, major hostilities of importance to the United States were about to break out somewhere in the world, Charlie could unilaterally issue a formal "Warning of War." It was not something one did lightly. Erroneously issuing that red alert could easily be a career ender. Yet, as the heat and humidity in Washington climbed that July morning in 1990, Allen began scribbling out his Warning, his message to the President informing him that Iraq was about to invade Kuwait.

For weeks he and his staff had been watching a war of words between the two countries. Only three years earlier, the Baghdad government of Saddam Hussein had ended the bloodiest war ever fought in the Middle East, the seven-year war between Iran and Iraq that ultimately left over a million people dead. Charlie Allen had followed the flow of that conflict in microscopic detail. Several times when it looked as though Iraqi forces would be surprised and overrun, the United States had passed Baghdad sensitive intelligence aiding Saddam, allowing him to move his forces into blocking positions. Never mind the fact that Iraq frequently stopped the swarms of Iranian volunteers with artillery barrages of chemical weapons, including sarin and mustard gas.

Both sides claimed that the other nation had started the war, but it seemed likely that it was really Saddam Hussein who had seen an opening after the fall of the Shah, using the opportunity to make a grab for the Arab-majority part of Iran. Conveniently, that region also held most of Iran's oil deposits. During the conflict, Iraq had relied heavily on Kuwait for financing the war, eventually to the tune of $14 billion. After seven years, with both sides on the verge of bankruptcy and collapse, the war ended with the original borders intact. That was 1987. It was now 1990.

Charlie Allen knew that Saddam Hussein still needed money to pay the Kuwaitis, or a way to coerce debt forgiveness from them. He also needed a political and/or military success. What he did not need were a bunch of royal sheikhs fooling around with the oil market, and that is exactly what Saddam thought Kuwait was doing. OPEC had agreed on a price for oil, as well as quotas for how much each nation could export. By violating that agreement, Kuwait was hurting Iraq's oil export revenue. Moreover, Baghdad also claimed that it had not been adequately compensated by the other Arab nations for all of its losses in the war against Iran. The Iraqi government portrayed its fight with the Persians as something it had done on behalf of all Arabs. Kuwait should be responsible for ponying up more of the tab by forgiving most, if not all of the debt Iraq had incurred during the war.

CIA's analysts of the Arab world thought that they had seen this kind of thing many times before, Arab capitals arguing with each other about their oil wealth. In the end, the Arab rulers usually reached an agreement, and the Western world would never hear of the controversy again. That would happen again in this instance, the CIA had been writing in its classified publications. Indeed, that is exactly what they had told President George H. W. Bush in the Agency's flagship publication, the National Intelligence Daily (NID), on July 25. They asked the question "Is Iraq Bluffing?" and answered it in the affirmative.

Allen had read the CIA's report and was deeply disturbed. He knew that it was wrong. The Iraqis were doing things they wouldn't do if it were only a bluff, but now Charlie had to find a way to prove it. He was intimately familiar with the NID. He had worked on the staff that put it together every day in the 1970s. Now, as NIO for warning, he had no say over what was in the President's daily digest of secrets. If he had, he would have been more cautious and would not have articulated the situation as practically guaranteed that Saddam Hussein's dispute with Kuwait about oil prices and quotas would come to naught.

Allen and his staff watched as the Iraqi Army began repositioning units to the south. "Saber rattling" was what CIA's experts on the Arab world had labeled the military movements. The Arabists in the State Department, almost all of whom had lived in Arab capitals for years, agreed with their CIA colleagues. They assessed it all as elaborate theatrics intended to send the message to Kuwait that Saddam meant what he said: he wanted some money or debt relief.

Charlie Allen and his staff, whom he later laughingly called "a bunch of misfits and iconoclasts," may not have been experts on the Arab world, but they were steeped in the literature and history of military science. They knew what military units to monitor and had practiced such analytical reconnaissance techniques extensively on East Germany, North Korea, and the Soviet Union. If a nation were really going to war, certain key military units would do things that they would otherwise almost never do.

During the Cold War, for example, the Soviet Union had crammed over a half million men and seven thousand tanks into East Germany just minutes from U.S. forces in West Berlin. In a few hours, they could be overwhelming U.S. and allied forces in West Germany unless the U.S. Army saw them coming and quickly took up defensive fighting positions. Thus it had been vital for over forty years that U.S. intelligence catch the initial movement of the Red Army to its jumping off positions for a war in Germany. Russian military exercises and maneuvers happened with great frequency, so knowing with confidence whether military movements were exercises or actual war preparations was critical. American intelligence did everything it could to tell the difference between the two, including a covert collection operation tunneling from West Berlin into the Soviet sector to tap the Red Army's military communications cables. A young Charlie Allen had been one of the U.S. intelligence officers who read the messages that were intercepted by that storied operation.

What the U.S. military had learned from watching the Red Army and North Korean military for decades was that there existed certain "tells," what in military parlance are called I&W: indications and warning. In a training exercise, for example, division commanders might not move all of their ammunition from storage bunkers toward the front, but in a real war they would. The problem was that the Soviets and the North Koreans tried to make their major exercises as realistic as possible to catch any defects in their system. Some Americans also thought that the Soviets and Koreans knew what organizations the Americans looked at for warning and would exercise those units too, just to make it more difficult for us to tell if there were a real attack being planned. It was a dangerous game.

If the "other side" mistook an exercise for war preparations, it might decide to go on alert, or perhaps launch a preemptive attack. This had, in fact, almost happened. Reacting to the harsh anti-Communist rhetoric of the new Reagan administration, the Soviet leadership had discussed in 1981 whether the new President intended to start a war. Then in March 1983, the United States and its NATO allies staged the most realistic nuclear war exercise that NATO had ever performed. The Soviets, already looking for a surprise attack, thought the Able Archer exercise was cover for a real attack. They ordered the Red Army to nuclear alert, sending submarines to sea, putting bombers in the air, and fueling missiles. When the NATO exercise ended, the Soviet Union gradually backed down too, but the incident brought the United States and Soviet Union closer to war than they had been since the Cuban Missile Crisis twenty years earlier.

Certain tells had been identified by Charlie Allen and his staff to monitor the situation between Iraq and Kuwait. Months before, in January, he ordered his staff to review the Iraqi I&W target list. They refined the list and increased the priority of photographing the key units. He also ensured that the photo interpreters would give greater priority to those images.

The tells Charlie was looking for were revealed in satellite pictures taken in late July: the Iraqi military was rolling out all of the support units that would assist combat units in a full-scale war. These were units that did not normally participate in field exercises and would not be needed for a limited operation. They were also moving ammunition out of storage and trucking it south. This was no drill. Allen knew that none of that activity was necessary to scare the Kuwaitis. The Kuwaitis couldn't see it. They didn't even know it was going on, but Charlie did.

He formally issued that judgment in the Warning of War notice and, using his secure fax machine, sent it to the White House Situation Room. He had not cleared it with anyone in the CIA or even with his nominal boss, the chairman of the National Intelligence Council. They would just have slowed it down, watered it down. There was no time left for that.

WHAT IF THEY HAD LISTENED?

As Charlie Allen now remembers the reaction of his CIA colleagues, "the wolves were out to shred me."[1] He was seventy-nine years old when we met to talk about the episode, and despite this story taking place over twenty-five years earlier, his memory was as encyclopedic and precise in our meeting as ever. He recalled specific dates, intelligence reports, words used in memos from decades earlier. "They threatened to fire me." It was neither the first, nor the last time that possibility would pass through the minds of Allen's CIA supervisors. "Fritz Ermarth, the NIC Chairman, thought I should have cleared the warning with him. Dick Kerr, the Deputy Director of CIA, was irate that the intelligence community was telling the President two different things." Charlie Allen was not intimidated. "I had the authority to issue a Warning of War and I did it." He smiled with a knowing look and added, "Moreover, I was right."

Few thought so on July 26. The CIA's response to the Warning of War memorandum acknowledged that Iraq might stage a limited operation, perhaps occupying an island owned by Kuwait in the Persian (or Arab) Gulf. They called it the Bubiyan Grab option. Or, they admitted, Iraqi forces could possibly drift across Kuwait's northern border in the vast desert, seizing areas that contained Kuwaiti oil fields. These possibilities were still unlikely and, if they happened, would be temporary moves to force Kuwait's hand in the talks regarding money and oil.

The Arab experts in the State Department agreed. "No Arab nation has ever gone to war with another Arab nation. It does not happen," they told Dick Clarke (then the Assistant Secretary of State for Politico-Military Affairs), repeating the conventional wisdom. "Besides, it's too hot. Temperatures in the Kuwaiti desert in late July are above 120 degrees Fahrenheit." Most Arab leaders left the region that time of year for Switzerland or the south of France. Washington's national security leadership was also packing up for cooler climes, like Maine. Washington always shut down in August, and August was just a few days away. No one wanted a crisis that would force them to cancel a vacation.

The Arab leaders contacted by the State Department downplayed the threat from Saddam. King Hussein of Jordan flatly rejected the possibility of his neighbor invading Kuwait. The exception was Sheikh Zayed of Abu Dhabi, who had been warning Washington for weeks that Saddam was planning to go to war. Zayed, the founder and President of the United Arab Emirates, wanted Washington to deter Saddam, to militarily signal that the U.S. would not tolerate any further Iraqi aggression. The UAE asked for American aerial refueling tankers, KC-135s, to be deployed quickly to Abu Dhabi so that fighter jets operating out of the UAE could fly round-trip missions up the Gulf to Kuwait, but the Pentagon was not interested. They had just held a naval exercise in the Gulf anyway, what more signaling could be necessary? In fact, U.S. ships were already leaving.

Several days passed in which the U.S. could have signaled Saddam, by either sending the KC-135s, or more directly by sending a high-level envoy from Washington. President Bush was already well known for the hours that he spent on the phone personally chatting up leaders of nations throughout the world. It would not have been out of the ordinary for him simply to have called Saddam. Saddam had frequently met with high-level U.S. delegations, including ones led by former defense secretary Donald Rumsfeld and Republican Senate leader Bob Dole. Making contact with Saddam via any of these channels could have very well prevented the entire Gulf War, yet the Bush administration made no call, sent no envoy. The leaders didn't believe that the situation was as dire as Charlie had suggested.

Six days after the Warning of War notice, on August 1, satellite images showed that Iraqi Republican Guard divisions had moved into attack formation in the desert very close to the Kuwaiti border. Charlie Allen started working the phones, calling urgently those from the White House and the State Department who he thought were still in town. He wanted the most senior national security group, the Principals Committee, to meet. Few of them were in Washington. He urged a meeting of the next most senior group, the Deputies Committee. Its chairman, Deputy National Security Advisor Robert Gates, was on vacation. Finally, the Undersecretary of State, Bob Kimmit, agreed to host a Deputies Committee–like meeting at the State Department.

They assembled in a wood-paneled conference room near the Secretary of State's office on the afternoon of August 1. The CIA, represented by its Deputy Director Dick Kerr, allowed that it was possible that Iraq might temporarily occupy some oil fields near the border, maybe even attempt the Bubiyan Grab. Kimmit then asked Allen to make his case that a full invasion and occupation was about to happen.

Charlie Allen laid out the details: which military units were where, the tells indicating that this was no exercise, and the fact that almost all of Iraq's Special Forces were located at their helicopters or commando boats. Then, in his quick, precise, but monotone style, he added "And most of the units in attack position have stopped communicating by radio. They have gone into EMCON—emissions control—they've gone silent." At that point, both Dick Clarke from State and Richard Haass from the White House looked up from their note taking, staring across the table at each other in shock. Both of them students of military history, they knew what it meant when an army went silent. Clarke silently mouthed his reaction to Haass, "Oh, fuck." Haass nodded in agreement. Still, not everyone was convinced, and the meeting adjourned without an agreement to do very much.

Only a few hours later, in the evening, the State Department Operations Center called all participants, urgently asking them to reconvene. People at the U.S. Embassy in Kuwait City were reporting that they could see, from their windows, Iraqi Special Forces landing in helicopters. A short time later, they reported that an Iraqi tank had parked in front of the building. When the meeting reassembled at State, Bob Kimmit began by dryly noting, "Well, it seems that Saddam has done more than grab Bubiyan." The Iraqi military had completely occupied the entirety of Kuwait in a matter of hours.

Within a week, George H. W. Bush would order a huge U.S. military force into the Persian Gulf, and within a few months that force would grow to over half a million men and women. What followed was the Gulf War, now known as the First Gulf War because there has since been a second. American military involvement in Iraq still continues, now in its twenty-fifth year.

What could have happened if decision makers had listened to Charlie Allen and taken seriously his Warning of War? It is, of course, impossible to know. Alternative history is a parlor game. Moreover, Saddam may have been intent on making Kuwait the eleventh province of Iraq no matter what was said to him or the degree of U.S. intervention, but the possibility remains that he may have been deterred at a far lower cost than the eventual outcome.

PRECEDENT WAS PART OF THE DECISION

Charles Duelfer, who led the UN (and later, CIA) efforts to deal with Iraq and its alleged program to build and maintain weapons of mass destruction, thinks that the First Gulf War was a colossal case of mis-communication. Duelfer, who wrote many of the questions asked of Saddam Hussein in captivity and later analyzed his answers, believes that the Iraqi leader had no idea that the U.S. would object so strenuously to his land grab. Had he thought that the U.S. would deploy a huge army to eject him from Kuwait, he never would have invaded. The problem was that on August 1, 1990, even the U.S. itself didn't know that it cared enough about an Iraqi seizure of Kuwait that it would act militarily. That government policy emerged only gradually in the months that followed.

Whatever might have been, we do know enough about what happened to believe that Charlie Allen qualifies as the Cassandra of the First Gulf War. He loudly and explicitly warned that it was going to happen, he was doubted and vilified, no one acted on the basis of his prediction, and he turned out to be right. Allen is, however, a different kind of Cassandra than many of the others we examine in this book. The job that he held was that of an institutionalized Cassandra, one created because of painful past mistakes, such as Pearl Harbor. His case is all the more disappointing because even the warnings of an official, full-time Cassandra were not heeded. Why?

The first answer involves personality. Charlie admits that many in the CIA considered him and his staff to be "a burr under their saddle." He calls it having sharp elbows, and his critics called him abrasive. Clearly, he was different from his colleagues. One of the reasons he stood out was his style and work ethic. He always seemed to be the first one into the office in the morning and the last one to leave at night. He was more experienced than all but a handful of senior officers, having served in the CIA and watched its successes and failures from the inside for decades. Meeting Charlie Allen, you know you're dealing with someone unlike the normal intellocrat. With his deep, sometimes booming voice, he speaks in precise, rapid sentences filled with references to dates, places, people, intelligence reports, and collection technologies. You can almost see the footnotes whizzing by in a blur of data. He speaks with the conviction of someone whose thoughts have been steeped in deep analysis, seldom opining on matters unless he has spent time considering them in depth using ample research material.

Many bureaucrats found that disposition intimidating. He was simply better at arguing his point of view than others in the U.S. intelligence community. He could overwhelm you with facts you didn't know. Seeing Charlie coming, one colleague told us, some people saw a bulldozer, a force of nature. Rather than reacting by thinking that Allen had painstakingly collected and pieced together a comprehensive set of information to come to his conclusions, some reacted to his style by concluding that he was just trying to garner attention for whatever project he was assigned to work on, presumably at the expense of other projects that possessed a justifiably higher priority.

Part of his reputation was also that he had been "controversial." That label actually meant that Allen had never shied away from issues or problems that more cautious bureaucrats would see as career threatening. When a CIA station chief had been captured and was being tortured by Iranian-backed terrorists in Lebanon, Allen and some in the Reagan White House tried to create a covert opening with Iran, secretly traveling to Tehran. The efforts ultimately led to the Iran-Contra scandal (after money began changing hands) and the indictment of senior Reagan administration officials. When it became public, the CIA formally reprimanded Charlie Allen for his role. He fought back, pointing out that it was he who had blown the whistle on the illegality of the money transfers involved. Defended in a CIA hearing by Washington lawyer Jim Woolsey (who would later become CIA Director), Allen's reprimand was withdrawn. Nonetheless, the episode added to the image of a man who was different in some way, perhaps reckless, perhaps dangerous to the bureaucratic establishment.

A second reason why Allen wasn't heeded by his superiors was organizational. Because it was the warning officer's job to draw attention to matters of immediate concern, some believed that the office would issue warnings simply for the sake of bureaucratic survival. Fifty years after the disaster at Pearl Harbor, which had led to its creation, some people in the intelligence community had forgotten why the Warning Office even existed. Even though Allen was cognizant of the need to avoid becoming "the boy who cried wolf," some in the intelligence community assumed crying wolf was what the Warning Office was all about. This view was usually accompanied by a concomitant disdain for people without extensive regional expertise.

The Middle East office knew more about its area, some assumed, than the analysts in the Warning Office, whose knowledge was "a mile wide and an inch deep." Charlie Allen was aware that the kind of people who gravitated to his staff were often those who didn't fit in well within traditional bureaucratic settings. He valued that quality. Others saw the team as a bunch of misfits. People in the CIA were accustomed to analysts with regional expertise. They didn't always understand that warning had become a sophisticated functional discipline and that it was the Allen misfits who had had mastered it. Charlie called it "seeing things through a different prism."

Dick Clarke found a similar problem while managing a team of about a hundred intelligence analysts in the State Department. There was a deeply held prejudice in favor of the views of the regional expert who had been on the desk for a long time. Long tenure certainly provided valuable insights, but it also made it difficult for the analyst to detect when real change occurred. When presented with indicators that something new was happening, that a situation might have reached a turning point, the seasoned experts would often recount the numerous times similar indicators had been seen before and had resulted in no change. They had seen many false dawns, and seeing them had resulted in a resistance that blinded them to small yet important differences from past events. They would tell Clarke that people brought in as a "set of fresh eyes" (i.e., him) always thought that something new and important was about to happen just because they lacked experience on the file. These analysts who had been on the accounts for a long time sometimes couldn't tell the difference between past events that hadn't resulted in a crisis and those causing the real crisis when it came along.

This leads us to the third element in the Iraq-Kuwait case, a problem we will encounter time and again in this book: lack of precedent, what we will refer to as Initial Occurrence Syndrome. "No Arab nation has ever gone to war with another Arab nation, or invaded a brother Arab's territory." When the regional analysts said that, left unsaid was "and, therefore, it will never happen." Had they explicitly made that claim, it might have been easier to see they had made a hasty generalization based on insufficient evidence. Instead, it was left implicit that if a phenomenon had never happened before, it never would. As illogical as that belief usually is, it has proven a powerful determinant in the way people react to Cassandras.

Psychologists have a term for this phenomenon: the availability bias. It is one of a series of cognitive biases, which are shortcuts the brain uses to make decisions quickly, processes that were created and refined over the millennia of human evolution to improve survival. And for cavemen, they did. When our ancestors had to decide whether or not to explore a cave they'd never seen before, their brains rapidly accounted for how often they'd experienced threats lurking in caves and made a rapid probability calculation as to whether or not exploring the cave was a good idea. Cognitive biases worked well when rapid pattern recognition and decision making was critical for survival, but in the modern world these processes can have unintended consequences.

The availability bias tends to prejudice our interpretation and understanding of the world in favor of information that is most accessible in our memory, things that we have experienced in the recent past. Think back in your own experience: when have you been more hesitant to take a flight because of the threat of terrorism? Two weeks after the September 11 attacks? Or last Tuesday? At neither time were there elevated indicators that a terrorist planned to hijack an airliner, but in the weeks following 9/11, anxiety over the safety of air travel was at an all-time high. The availability bias is even less helpful when trying to think about and mitigate future disasters.[2] We consider Initial Occurrence Syndrome a special case of availability bias, one that is more difficult to overcome because of the complete lack of precedent that would allow our brains to estimate the likelihood of such an event occurring.

For Allen, however, precedent was part of his decision to sound the alarm. As he looked at the indicators of war in the buildup of the Iraqi military in July 1990, he admits that he was haunted by a memory of a past mistake. One morning in October 1973, while Charlie was assigned to the staff that created the President's Daily Briefing, word came in from the National Security Agency that the Egyptian military seemed to be going on high alert. Charlie was not the warning officer then, and he assumed someone else would analyze the problem. He took the elevator down to the car waiting in the CIA parking lot, omitting the information from the briefing. Hours later, the Israeli Defense Force was hit by a surprise attack, resulting in a number of casualties and a near defeat. Both Washington and Tel Aviv were caught flat-footed.

Seventeen years later, the 1973 Arab-Israeli War served as a helpful and personal reminder to Allen that surprises do happen. Could this event have also become a source of guilt that distorted his analysis of Iraq-Kuwait, driving him to err on the side of warning? Personal experiences are often part of what drives Cassandras to make the choices they make. The late Harvard historian Samuel Huntington proposed that, in analyzing leaders, it is always good to know what world events and personal experiences shaped them while they were young and their world view was being formed. Both the Cassandras and those evaluating their claims should be cognizant of the origins of the prisms they look through to see the world.

A fourth aspect of this case that often reappears with Cassandras is the role of empirical data. In the classical myth, Cassandra had visions. Claiming visions as your evidentiary source is, as the hexed young lady of myth discovered, unlikely to persuade many people. Modern-day Cassandras, however, seldom claim magically derived insights. Instead, they often seek out and interpret data that others overlook, collect facts that no one else bothered to assemble, or derive new insights from data others already have. The National Warning Officer had been given an ability to order collection of data. He could move satellites, prioritize the processing of intercepted communications, and task spies with key pieces of information to seek out. When Charlie Allen thought about the Iraqi military buildup, he was looking at a set of objective indicators that he and his staff had drawn up months before the crisis. What he saw was that a long list of empirical warning indicators had moved from SAFE to DANGER.

Unlike Charlie and his team, the regional analysts were not particularly focused on these indicators, nor did they consider them valuable. When Charlie Allen said that the ammunition depot in some small Iraqi town was empty, he expected alarm bells to go off in the minds of his audience, but many in his audience had no idea whether such a fact was important or not. Across all fields we have studied, expert Cassandras often make appeals to the compelling nature of data they themselves collected or which they endorse as important. The data speaks for itself, the Cassandras believe. For non-experts, however, obscure data or variables that they have never heard of before do not speak to them in the same way, and they understand them no more than they would a sentence in a language they don't know.

Finally, the case of Charlie Allen also proves that being right isn't always rewarded. Following Iraq's invasion of Kuwait, he was promoted to Assistant Director of Central Intelligence for Collection, the person who prioritizes the targets for U.S. satellites and spies. Almost a decade later, in the months after the 2003 U.S. invasion of Iraq, Allen traveled to that country to smell around, to see what he might be missing while cloistered away in the Washington bureaucracy. Not surprisingly, he found data that others had not. He found trends that were disturbing, on the verge of alarming. He found what he believed were the elements of an incipient insurgency, one that could become quite powerful and one for which the U.S. occupation force was not prepared.

Upon returning to Washington, Allen reported his findings in a personal memorandum to the new CIA Director, Porter Goss. Goss had been a longtime congressman and was not entirely proficient in the ways of the executive branch or the intelligence community, yet he saw immediately the significance of Allen's report. He handed it to the President, and the President actually read it. Charlie Allen had not anticipated either of those moves. President George W. Bush then asked his new Director of National Intelligence, John Negroponte, what he thought of Charlie Allen's report on the threat of insurgency in Iraq. Negroponte was blindsided and embarrassed. He was also irate at what he believed was an attempt to circumvent his authority and report unilaterally to the President.

Negroponte soon took his revenge. During a bureaucratic shakeup of the intelligence community, Charlie Allen's position was shifted from the CIA to the office led by Negroponte. Everyone assumed that Charlie would continue to lead it, but Negroponte said he wanted someone else. Left out of the reorganization, Allen was offered the dubious opportunity of a transfer to the problematic Department of Homeland Security (DHS). Long past retirement age, but ever the devoted civil servant, Charlie accepted the challenge. He then set about trying to create an intelligence collection and analysis capability covering the seventeen agencies that had been recently mushed together to create the DHS.

When he did ultimately retire, after fifty years in the intelligence community, Charles E. Allen was given a host of awards and accolades by the many agencies that make up the U.S. national security and intelligence apparatus. He was toasted at formal dinners of retired spies. Ironically, many articles were written in the media about him, although he always treasured his obscurity. A chorus of past CIA directors sang his fame and praised him for all that he had done to warn the country about Iraq, in addition to so many other crises.

In 2009, the position of National Intelligence Officer for warning was abolished. In explaining the decision, one official noted, "It's the job of all the national intelligence officers to warn." Perhaps so, but the explanation raises a question: who among them is actually listening for Cassandra?

CHAPTER 3

The Rebuilder: Hurricane Katrina

There is no faith like the faith of a builder of homes in coastal Louisiana.

—DAVE EGGERS, *ZEITOUN*

fter gaining strength during its relentless march through the Caribbean, the massive Category 3 hurricane swung north toward Louisiana. The small coastal communities in Plaquemines Parish and on Grand Isle would be the first to fall victim to the pounding rain, shearing winds, and most dangerous, the massive, fifteen-foot storm surge. Many longtime residents of the bayou had successfully ridden out past storms and refused to leave; even if they changed their minds at the last minute, the few roads heading north were already jammed with traffic. Barely further inland, the city of New Orleans, the Crescent City, the Big Easy, one of the cultural gems of the United States, was nearly as vulnerable. As the slow-moving storm made landfall over southeast Louisiana, emergency response officials huddling in the Louisiana State Emergency Operations Center braced for the worst. The looming catastrophic impact of the hurricane was beyond question. Ironically, the network of drainage canals that slice through New Orleans and are instrumental in pumping storm water into Lake Pontchartrain, just to the north, would help funnel the storm surge right into the heart of the city. Only the series of decades-old levees, flood-walls, canals, and pumps, constructed by the Army Corps of Engineers, stood between metropolitan New Orleans and complete devastation.

But the levees, never built to withstand a storm of this magnitude, were woefully inadequate. As the hurricane tracked slowly west past New Orleans, the storm surge flooded into Lake Pontchartrain, easily overtopping the levees. Within hours, the damage to metropolitan New Orleans was catastrophic. In addition to the rising floodwa-ters, twenty inches of rain and 120-mph winds leveled buildings and homes, rendered major roads and canals unrecognizable as well as impassable, inundated hospitals with both victims and water, knocked out power and phone service, and flooded the region with sewage and hazardous chemicals. The human cost: 175,000 injured, 200,000 sickened, and at least 60,000 dead, many of whom were simply un-able to evacuate because they were too old, sick, or poor to arrange reliable transportation out of New Orleans before disaster struck.

This was the hurricane New Orleans had long feared, but it was not Hurricane Katrina. It was a disaster-preparedness simulation called Hurricane Pam run by the U.S. Federal Emergency Management Agency (FEMA) in late July 2004, just over thirteen months before Katrina made landfall over the same part of the country. The exercise combined the expertise of some of Louisiana's foremost hur-ricane and disaster experts with emergency management officials from the federal, state, and local levels. The Cassandras hoped, after decades of halfhearted fixes punctuated by willful neglect, that their warnings to seriously plan for and mitigate the hurricane threat to New Orleans had finally been heeded. Tragically they learned, barely a year later, they were wrong.[1]

SINKING INTO THE WAVES

The threat from hurricanes and tropical storms and their associated flooding is soaked into the rhythm of life in the Crescent City. Its residents expect a solid deluge in the summer months with almost nonchalant familiarity, like seeing king cakes at Rouses before Mardi Gras or watching a line of music lovers following a brass band through the streets on a sweltering summer Sunday. Storms have plagued New Orleans since its establishment. In 1722, only four years after its founding by the French-Canadian Sieur de Bienville, the settlement was effectively destroyed by an eight-to-ten-foot storm surge during a hurricane.[2] The unique topography of New Orleans, and indeed the entire Mississippi Delta Region, makes it particularly susceptible to extreme weather and has bedeviled efforts to mitigate nature's wrath.[3]

Dr. Ivor van Heerden has studied wetlands and river mouths since 1969, first in his native South Africa, and later as a Ph.D. student at Louisiana State University (LSU). Van Heerden is intimately familiar with Louisiana's coastal topography and the dynamics of the Mississippi Delta. As he describes it, "just about the whole state of Louisiana south of Interstate 10 is new land, relatively speaking . . . laid down over the five millennia from the almost annual spring floods of the [Mississippi River]." This sediment is watery, but it is also filled with organic matter. The entire region is sinking dramatically, about a couple of feet per century, as the organic matter decomposes and the soil is compressed under its own weight. For over a century, this sinkage didn't cause any major issues, as settlers in the area built their own small, ad hoc levee systems to divert flood waters and protect themselves from the Mississippi River.

That changed in 1927. After months of heavy rain, the Mississippi overflowed its embankments throughout the Midwest and destroyed levees, flooding over 27,000 square miles, killing over two hundred, and causing $5 billion in property damage (in today's dollars). It was the worst flood in American history.[4] Congress subsequently passed the Flood Control Act of 1928, directing the Army Corps of Engineers to build the system of levees, floodways, and drainage basins that continue to control and contain the flow of the Mississippi River to this day.[5] But this labyrinth of engineering has an unintended consequence: new sediment is no longer laid down over the old, resulting in ubiquitous potholes, as any despairing Crescent City car owner can attest. In fact, the entire region is literally sinking farther and farther into the waves; not the best news when nearly 50 percent of the city already lies below sea level.[6]

Complicating matters further is the dramatic loss of coastal wetlands over the last several decades, wetlands that provide a natural barrier to erosion and storm surge. The Gulf Coast is the source of most of the nation's oil and much of its natural gas, making it an indispensable part of the Gulf states' economies.[7] Most people are familiar with oil companies' offshore oil drilling operations, but many don't realize that companies also drill directly into and beneath the Gulf wetlands to tap vast reservoirs of hydrocarbons, construction operations that permanently damage and destroy the delicate marsh ecosystem. Moreover, transporting all of this oil and gas requires thousands of miles of canals and pipelines, dredged and carved out of the precious wetlands. The result was a net loss of over a million acres of wetlands between 1930 and 2005. Dr. van Heerden pointed out that more than half of this lost wetland once lay directly seaward of New Orleans and, in the event of a hurricane, "would be sapping the strength of the storm winds and quite literally swallowing the brunt of the storm surges."

After finishing his graduate studies at LSU in 1983, van Heerden returned to his native South Africa, where he worked on coastal management before fleeing the country's political turmoil. Somewhat listless, he eventually returned to Louisiana at the urging of a former colleague, first working as a consultant and later setting up LSU's Natural Systems Management and Engineering Program, focusing on coastal restoration and preservation of the wetlands he knew so well. At the time, Louisiana already had a federally funded program to protect and restore its wetlands. Despite its earnest name, the Coastal Wetlands Planning, Protection, and Restoration Act (awkwardly referenced as CWPPRA, or "quipra") was fundamentally flawed: "They weren't thinking big enough," as Dr. van Heerden recalled.[8]

From the start, he knew that Louisiana needed comprehensive coastal restoration specifically because of the hurricane threat. "The first thing I did was to write a plan, a big-picture plan," Dr. van Heerden told us. The plan got a lot of media coverage and lofted him to rock-star status, at least among the officials and scientists that were thinking about Louisiana's marshes. One of those officials happened to be Governor Edwin Edwards, who appointed him assistant secretary of the Department of Natural Resources in 1994. As the state's "restoration czar," his express responsibilities included engineering short- and long-term solutions to the state's coastal health issues.

Finally he had the mandate and authority to mitigate the disaster he feared was coming. He would start with rebuilding the barrier islands, offshore shoals of sand that act like the wetlands in blunting the effects of hurricane storm surge, then follow by reclaiming and rebuilding the disappearing coastal wetlands. Unfortunately, however, pet projects and special interests got in the way of preventing the preventable disaster. Local politicians introduced a bill that effectively gutted the state's broader efforts to regenerate the coastal wetlands by forcing the government to compensate property owners for *any potential* economic losses due to the environmental reclamation efforts, an attempt to curry favor with well-heeled local landowners.

Ivor van Heerden is, by all accounts, an outspoken scientist. To his scientific mind, these political machinations were an attack on everything that made sense. Decades of scientific investigation and empirical evidence proved that humans were causing Louisiana to lose coastline at an alarming rate. The coastline, which appeared to some as useless swamp, was inevitably the key to protecting all that Louisianans had built in the preceding centuries. Without a serious effort to restore these defenses, New Orleans was one major storm away from being underwater.

He and his colleagues went on a full-throated defense of their plan, writing their congressmen and rallying public support, but in the end it was not enough. After the next election, and before any of the projects had even gotten off the ground, the restoration efforts were killed, CWPPRA was neutered, and van Heerden's position and influence as the state's restoration czar looked ever more tenuous. Frustrated by the backroom dealings of Louisiana politics, van Heerden resigned in disgust.

HIGH TECH, HIGH STAKES

Dr. van Heerden knew the danger only grew with each passing day, and he soon found a new authoritative platform from which to continue sounding the alarm. After leaving government, he was rehired by LSU. In 1998, the university established the LSU Hurricane Center, a virtual organization that brought together scientists to study and model the storm surge, winds, and flooding that would result when a future, major hurricane made landfall over southeastern Louisiana. Marc Levitan, then a professor of civil engineering at LSU, was named director; Ivor van Heerden became deputy director. "Once we had the hurricane center, it gave me the vehicle to go out and get the money to do what I really wanted to do: to take a hard look at New

Orleans, develop storm surge [computer] models, and really understand what was going on."

The team soon realized that studying the second- and third-order impacts of such a storm required funding that surpassed what the LSU Hurricane Center was receiving. So, after concerted lobbying efforts and funding requests, LSU established a second institute in 2002, the Hurricane Public Health Center, with Dr. van Heerden at the helm. Modeling the storm surge and its effects on the city of New Orleans was of a critical importance that was recognized well outside the confines of LSU. "We managed to convince the governor that LSU needed a supercomputer," van Heerden recalled with a smile, "because we *needed* the supercomputer."

The hurricane threat to New Orleans had long been familiar to emergency preparedness officials. There's even a name for it in the community: the New Orleans Scenario. Despite decades of concern, no organization had attempted to comprehensively evaluate all the facets of the monumental disaster that would result from a direct hit by a slow-moving hurricane on the city of New Orleans. The LSU Hurricane Center and the LSU Hurricane Public Health Center did precisely that. Fully equipped with the brainiest of scientists and graduate students, the team set about modeling hurricane storm surges and wind patterns, developing flood predictions for the Greater New Orleans metropolitan area. The teams eventually went even further, studying the resilience of the city's flood-protection system, problems that would result from a mass-evacuation order, the pollution and contamination that would result when rising storm waters mixed with the contents of nearby industrial warehouses and chemical plants, and the infectious diseases that could rapidly spread through the marooned population once floodwaters subsided.[9]

The team was troubled too by the quality and height of the levees surrounding the Crescent City. The Army Corps of Engineers assured van Heerden and his colleagues that there was no cause for concern, that the levees were built to withstand the "most severe combination of meteorological conditions expected." That was the language Congress used when it passed the Flood Control Act of 1965, directing the Corps to build and modernize New Orleans's flood-protection system after large portions of the city were flooded by Hurricane Betsy. The scientists' models suggested the current levees were not tall enough to withstand the storm surge that could reasonably be expected from a Category 3 hurricane. They worried further that the soft clay on which the levees and floodwalls were built would undermine their structural integrity should they ever be tested to their limits; concerns that later proved to be prescient.[10]

Still, Cassandra finally seemed to have an influential platform to spread her message; and spread her message she did. Dr. van Heerden and his colleagues published their research widely, sending their results to the media, local and state emergency-management organizations, hospitals, FEMA, and many other targets. Members of the LSU Hurricane Center were featured in local and national news stories. One favored stunt included taking news crews down to the French Quarter and holding interviews on roofs to show the crews and their viewers how high one would have to climb to stay above the floodwaters.[11] Maybe if the team couldn't get local officials to implement a long-term plan to protect and restore the region's wetlands, they could at least get people to plan for the inevitable catastrophe. "Every opportunity I had to talk, from the old ladies' knitting group to the mayor of New Orleans's environmental forum to the media, I spoke about it."

By 2004, van Heerden's relentless efforts appeared to be paying off. After several years of fits and starts, of funding problems and contract snags, FEMA would organize and hold a massive, comprehensive emergency-management exercise for the New Orleans Scenario.

Known as Hurricane Pam, the exercise would bring together experts and managers from federal, state, and local organizations, all of whom held a stake in seeing New Orleans prepared and reconstituted in the event that the worst happened.

The initial effort had actually been set in motion back in 1999, when Dr. van Heerden worked with the Louisiana Office of Homeland Security and Emergency Preparedness to request federal funding to hold an exercise to bolster the state's hurricane response. That request languished due to a lack of funding, a demoralized agency, and a nation singularly focused on the threat of terrorism. The George W. Bush administration saw FEMA as a bloated bureaucracy and, during the major government reorganization that occurred after the September 11 terrorist attacks, subordinated FEMA under the Department of Homeland Security, no longer reporting directly to the White House.[12]

Nevertheless, in 2004 planning was finally underway. Curiously, despite having already developed highly accurate computer simulations of hurricane winds and storm surge for the city of New Orleans, the team at the LSU Hurricane Center was not asked to supply the model for the exercise. Instead, FEMA awarded the contract to develop the scenario to an outside contractor, asking Dr. van Heerden and his team only to provide them with inputs and produce storm-surge animation. Still, he thought, something was better than nothing.

It was not better by much. He recalled during the exercise "federal officials and the Corps of Engineers officials would just scoff at us." As the Hurricane Pam exercise kicked off in the sweltering heat and humidity of late July, a sinking feeling in van Heerden's stomach began to rumble, and then it grew. In spite of his previous enthusiasm for mounting such planning efforts within the New Orleans emergency preparedness community, van Heerden was soon exposed to the unfortunate truth: while local officials maximized the opportunity to "war-game" their emergency plans, federal managers in town from FEMA didn't seem to take seriously the enormity of the probable catastrophe.

He made the rounds, still not clearly understanding why federal officials did not seem to fully grasp the gravity of the crisis, nor why they seemed to ignore the years of work and research produced by the LSU Hurricane Center and the Hurricane Public Health Center. The scientists had already foreseen problems with evacuating the population, particularly those too old, sick, or poor to find their own means of transportation out of the city. They foresaw the public health issues that would erupt following the catastrophic devastation of a major population center. They knew the logistical requirements for providing shelter, food, water, and medicine to stranded populations. "I tried to get FEMA to at least recognize the need for tents." "Tents?" one incredulous FEMA official exclaimed at Dr. van Heerden's suggestion. "Americans don't live in tents!"

Even if federal officials didn't seem to take the exercise very seriously, local officials refined their evacuation planning, search and rescue coordination, and emergency communications plans. Van Heerden credits the much-lauded success of the Coast Guard and Wildlife and Fisheries search-and-rescue efforts during actual Hurricane Katrina recovery operations to their assiduous involvement in the Hurricane Pam exercise.

Still, major aspects of the disaster planning went unfinished. The Hurricane Center had previously estimated, after conducting an extensive survey of the community and looking at data from past hurricane evacuations, that upwards of a hundred thousand people would remain in metropolitan New Orleans even after a mandatory evacuation order had been given. In the Hurricane Pam exercise, the final answer to the evacuation transportation problem was TBD.

Unable to marshal funding for more exercises to resolve such unanswered questions, the New Orleans Scenario disaster plan was never actually completed. Eventually, in the summer of 2005, two follow-up workshops were held, one focusing on transportation and the other on medical support. But it was too little, too late. By late August, Katrina was already churning through the Gulf of Mexico on its inexorable charge toward the city. Two days before the storm made landfall, pamphlets enumerating lessons learned and a draft transportation plan were hastily produced and distributed to emergency officials.

Years of warnings and advocacy, first for a long-term solution to Louisiana's crumbling coastal defenses and next for comprehensive disaster and recovery planning, finally collided with the dreaded inevitable. From the Louisiana Emergency Operations Center in Baton Rouge, Dr. Ivor van Heerden prayed for the best, but he knew that the region was unprepared for Hurricane Katrina as it made landfall over the Mississippi Delta on Monday, August 29. As reports trickled in over the subsequent hours that large swaths of the city had been inundated by rising salt water, Dr. van Heerden knew that New Orleans would never be the same.

Many subsequent investigations and reports focused on why the levees failed, why no comprehensive evacuation plan had been executed, and who should be held responsible for the botched recovery operation. The answers to those questions are not the aim of this book. The problems that befell the city of New Orleans during Hurricane Katrina had been discussed in the prescient warnings of Cassandras like Dr. Ivor van Heerden for years, even decades. Why had Cassandra's warning cries not resulted in action? Why had no leader stepped forward to mitigate the inevitable catastrophe?

While it is impossible to predict with high levels of precision exactly when and where a hurricane will make landfall, the New Orleans Scenario has long been considered one of the three most critical natural disaster scenarios facing the United States. From Hurricane Betsy in 1965 to Hurricane Georges in 1998 to Hurricane Isidore in 2002, New Orleans has suffered numerous near-miss hurricanes in just the last fifty years. Not only was the threat real, but the vulnerability and unpreparedness of the Gulf Coast region to deal with the destruction from a major hurricane were also well known. The *New Orleans Times-Picayune* published a Pulitzer Prize–winning series of articles in 2002 entitled "Washing Away," which ominously exposed, in painstaking detail, the precariousness of the continued existence of the Big Easy.

THE LEVEES FAILED, BUT YOUR GOVERNMENT FAILED TOO

In a sense, Hurricane Katrina is both the most straightforward and most tragic case study we examine in this book, because the affected communities and their leaders largely heeded their Cassandra's warnings, but their acknowledgement didn't translate into action that could have mitigated the impact of the inevitable storm. Why? We postulate that there were several primary reasons that Cassandra was effectively disregarded. The first was characteristics of the Cassandra himself; the second was biases inherent in those receiving the warning; the third was bureaucratic sclerosis; and last was the fact that the solutions flew in the face of the prevailing political winds.

First, because they are calling out disasters beyond our collective horizon, Cassandras are often seen as pessimistic, naysaying nuisances. This perception is amplified when an iconoclastic Cassandra becomes so convinced of the dire nature of her warnings that moral outrage drives her to starker and starker confrontation. Dr. Ivor van Heerden freely admits to wearing his heart on his sleeve and minces no words when asked about the many failures of leadership leading up to, during, and after Hurricane Katrina. His three-hundred-page narrative entitled *The Storm* lays bare his disdain for politicians who derailed efforts to rebuild Louisiana's wetlands, for FEMA officials who disregarded LSU's published disaster research and storm-surge models, and for politicians who thanked each other and congratulated themselves for acting decisively while thousands of people remained trapped in attics, stranded on rooftops, and marooned in the New Orleans Superdome. He recounts the naïveté of his early years, imagining that clear science and empirical evidence would open the way for bold action. He speaks with derision of the rapid efforts to deflect blame for the failure of the levees, the late and halfhearted evacuation plan, and the botched rescue-and-recovery response. Certainly this jaded perspective resulted from years of intransigent resistance to the warnings. These frustrations are understandable but are also not without consequence. Perhaps our Cassandras may learn from the old adage that you catch more flies with honey than with vinegar.

The second reason that Cassandra's warnings went unheeded has to do with the receivers of the warnings themselves. During a briefing to journalists on September 5, days after the levees failed and while the nation was still trying to grasp the magnitude of the disaster, Michael Chertoff, the director of the Department of Homeland Security, described the situation as, "that 'perfect storm' of a combination of catastrophes [that] exceeded the foresight of the planners, and maybe anybody's foresight."[13] His comment was widely mocked as indicative of the systemic ineptitude that pervaded the government's response, as experts like Dr. van Heerden had warned for years of just such a scenario (not to mention FEMA's own Hurricane Pam exercise that had concluded barely a year earlier). But it is also interesting because it illustrates a key observation that we saw with Charlie Allen in the preceding chapter and that we will see in Cassandra Events time and again: Initial Occurrence Syndrome is a potent obstacle to accurate evaluation of a warning.

One of the biggest hindrances Dr. van Heerden and his colleagues faced may have simply been that New Orleans had always scraped through before. Who was to say it wouldn't scrape through again? This seeming lack of imagination to accept that a more severe catastrophe could occur might be most easily summed up by the responses given by stranded bayou residents when asked why they didn't evacuate before Katrina: "We rode out hurricanes before; figured we'd just do it again." Recognition of this availability bias may have resulted in the diffidence of federal and local officials about genuinely preparing for the New Orleans Scenario.

The third factor complicating action in response to our Cassandra's warning is sclerosis within the organizations that could effect a solution. The Army Corps of Engineers, in particular, deserves particular blame for failing to heed warnings that were sounded for decades, and then moving too slowly to rectify the inadequacies while also failing to take any responsibility. New Orleans's modern flood protection system, as previously mentioned, was authorized by Congress in 1965, and was to consist of a series of levees and floodwalls that could withstand what was referred to as the standard project hurricane (SPH), a notional storm "based on the most severe combination of meteorological conditions considered reasonably characteristic of that region." In fairness, the SPH became a moving target. What was originally considered to be a "standard" hurricane for the region in 1965 was revised (and strengthened) over the years at least three times by the National Oceanic and Atmospheric Administration. Yet during this entire period, for what appears to be reasons only of organizational and bureaucratic expedience, the Corps failed to upgrade its designs or marshal the resources necessary to reinforce the city's hurricane protection system.[14] Even after Dr. van Heerden and his colleagues at the LSU Hurricane Center raised concerns that poor soil conditions and settling could cause structural and height deficiencies in sections of New Orleans's completed levee system, the Corps maintained that its designs were sufficient.

Moreover, project managers in the Army Corps of Engineers failed to heed the widely accepted and respected science supporting the need for restoration and protection of the coastal wetlands and barrier islands. Seeing such efforts as merely tangential to their main work of building walls to keep the water out, the Corps never seriously considered Cassandra's warnings of the increasing storm surge threat due to the loss of historically protective marshes and shoals. Even after the National Weather Service shared its concerns with the Corps that current efforts to protect New Orleans were possibly insufficient to protect the city from even a weaker Category 2 storm, no evidence exists that the Corps took any action to revise their designs or reassess their assumptions. While not necessarily malfeasance, bureaucratic ineptitude and decisions made for reasons of expedience seem to have diminished any serious regard for Cassandra's warning in this case.

The fourth and final reason our Cassandra's warnings were not heeded is, unfortunately, yet another theme repeated throughout this book: the solutions sought were not politically palatable and did not comport with the philosophy of the political administration at the time. Dr. van Heerden's initial efforts in the 1990s to develop and implement a coherent long-term coastal restoration program were stymied by short-term political gains sought by local politicians. When it came to his later efforts to encourage local, state, and especially federal officials to plan and prepare to mitigate a potential catastrophe, he discovered a similar problem. After a sobering experience in dealing with the aftermath of Hurricane Andrew in 1992, the Clinton administration set about reforming FEMA from the top down. A study by George Mason University later found that these reforms had, in fact, transformed the agency from "a bureaucratic, process-driven organization into a responsive, results-driven organization."

Unfortunately, seen as a bloated and irrelevant bureaucracy, FEMA was effectively gutted by the Bush administration beginning in 2001. Relationships among federal, state, and local offices atrophied as the agency's funding was slashed. Furthermore, after the September 11 terrorist attacks, counterterrorism efforts sucked up the funding, leadership attention, and perceived importance from dozens of other federal responsibilities. FEMA was subordinated to the newly created Department of Homeland Security, further removing the agency from relevance and causing morale to plummet. Michael Brown, the FEMA director at the time of Hurricane Katrina, had literally no emergency-management experience, having landed the job because of his connections to the Bush White House. Prior to joining FEMA, first as its general counsel and later as its deputy director and then director, Brown served as a commissioner for the International Arabian Horse Association. Such political and bureaucratic disregard helps to explain why FEMA officials were caught entirely off guard, with an organization that was underfunded and entirely unprepared to handle any disaster, let alone one on the order of Hurricane Katrina.

Sadly, many of these bureaucratic deficiencies and political motivations persisted through and after the Katrina crisis. Completed in 2015, New Orleans's new levee system was built by the Army Corps of Engineers to supposedly withstand a "1-in-100-year storm," approximately equivalent to Hurricane Katrina. But a 2013 study has already suggested that the levees in some parts of the city remain insufficient to withstand a Katrina-size storm surge. The Corps has since recommended that a complete review of its design be completed by 2018.[15]

Worse still was the outcome for Dr. van Heerden. Because he was recognized throughout the state for his tireless, decades-long advocacy to strengthen its coastal defenses, the Louisiana State secretary of the Department of Transportation and Development asked Dr. van Heerden to lead an independent forensic investigation into the causes of the levees' failures. True to his personality, the Team Louisiana Report, as it was colloquially known, does not mince words, squarely blaming the Army Corps of Engineers for its failure to build structures that met the basic requirements of the 1965 Congressional legislation. Despite findings that were broadly consistent with those of the four other commissions established to look into the Katrina disaster, the leadership at LSU, fearful that van Heerden's outspoken criticism of the Corps would cost the university federal research money, fired him in 2010.

Dr. van Heerden's warnings weren't welcome when they would have been most useful, and his attempts to establish accountability and ensure that past mistakes would never happen again were vindictively punished. After a three-year lawsuit, during which LSU spent nearly a million dollars in legal costs, the university abruptly abandoned its position and settled the case.[16] When we caught up with him in the late summer of 2015, Ivor was getting ready to head back to Baton Rouge for an LSU Hurricane Center reunion. Despite everything that's happened, van Heerden maintains the same spirit that drove his advocacy and warnings all those years: "I still feel like I've got a lot to give."

CHAPTER 4

The Arabist: The Rise of ISIS

We had been hopelessly labouring to plough waste lands; to make nationality grow in a place full of the certainty of God. . . . Among the tribes our creed could be only like the desert grass—a beautiful swift seeming of spring; which, after a day's heat, fell dusty.

—T. E. LAWRENCE, *SEVEN PILLARS OF WISDOM: A TRIUMPH*

The Arab Spring came in winter. It was mid-December 2010 when a final act of protest by a street vendor, protest by self-immolation, uncorked widespread, pent-up frustration against the aging dictator of Tunisia. Few people in Washington noticed. The Senate was closing for the holidays, leaving much work incomplete. On the list was President Obama's request to fill the long-vacant U.S. Ambassador's slot in Damascus. As soon as the senators left town, Obama used his right to make recess appointments, designating Robert Ford, the Foreign Service's leading expert on Arab affairs, the new U.S. Ambassador to Syria.

As Ford was unpacking in Damascus a month later, protests in Tunisia reached a crescendo, and President Ben Ali fled the country. Around the Arab world, people stopped in amazement and watched on television news what they would later call the Arab Spring. No one could seem to remember any Arab ruler being driven from office by street protests in recent history. The only long-serving Arab dictator recently run out of town was the one who had been expelled by American tanks almost eight years earlier. Could what just happened in Tunis occur elsewhere? That's what people in Damascus were asking as Ford made his introductory calls around town. Unspoken was the question: could it happen here? It seemed very unlikely, Ford heard people say. Tunisia was a one-off. That was what all the self-proclaimed experts said over little glasses of hot tea and demitasses of thick Arab coffee. Every Arab country is different from the others, after all. That is what Syrian President Assad told Ford directly, dismissing any parallel, any similarity.

Ford had been to most of the countries in the Arab world during his twenty-nine years in the U.S. Foreign Service. He previously served as U.S. Ambassador to Algeria and deputy ambassador to Iraq. He was not sure that those who saw Tunisia as an aberration were right. He was acutely aware of the frustration in most Arab countries, the perception in the "Arab street" that many of the ruling regimes could not successfully do anything to improve the living conditions and futures of their people. The numbers of disaffected were ever increasing. Many were young and unemployed. Ford had already served half a decade in U.S.-occupied Iraq, where he dealt with the violence and factionalism that had come rushing out of every nook when Saddam Hussein was deposed. Only now, eight years later, were U.S. forces getting to the point where they thought that the insurgents had finally begun to give up and accept the U.S. presence.

In those first few months of 2011, while Ford was settling into Damascus, protests began to form elsewhere around the Arab world, tentatively at first, then in unprecedented crowds, seen on television and followed on the new social media apps. By the time winter ended, the dictators in Egypt, Libya, and Yemen had been deposed. It seemed like a tsunami in the desert. The Spring was infectious. In March, protests began in Damascus.

For many in Syria, it seemed that it was time, time to protest against the rule of the Assad family and the minority Alawite sect that had gone on for over forty years. Maybe the Syrian dictator, like those in Yemen, Egypt, and Libya, could also be scared out of office by a few weeks of mass protest. Maybe the Americans and Europeans would bomb him, as they had done to Qaddafi in Libya. The rallies spread to other Syrian cities. In Hama, over three hundred thousand people took to the streets. Ford was not just watching; he was collecting information and intelligence, talking to everyone he could throughout the country, judging the depth and direction of the movement, and reporting back to Washington. He urged the State Department to make clear that the U.S. supported the peaceful protests. Washington agreed and authorized the U.S. Embassy to continue expanding contact with human rights groups and other elements of civil society, as well as opponents of the Assad regime.

The problem was that the most organized opponents were from the Muslim Brotherhood, which had been driven underground by brutal repression in the 1980s. Now, it began to emerge. The rest of the organized opposition to Assad had existed only in exile, in cafés in Paris and Istanbul, Dubai and Doha. As the spontaneous protests in Syria's major cities continued, the exiles watched. The usual Assad tactics, midnight arrests and shooting into crowds, were not deterring the opposition. Indeed, they seemed to fuel it. In city after city, large crowds clashed with the police, and in some places, with the army. Many anti-Assad exiles found ways to return and link up with the newly overt opposition cells in the major cities, both Muslim Brotherhood units and more moderate groups. There were also Syrian exiles in terrorist camps in places like Afghanistan and Iraq, but Assad's secret police had been effective for years in keeping al Qaeda out of the country or hidden well underground. Despite the Assad government having been one of the most reliable Arab partners the U.S. had in fighting al Qaeda before 9/11, an "al Qaeda in Syria" group existed, better known as the al Nusra Front. In the spring of 2011, it saw an opportunity.

Al Qaeda's central command, hidden in Pakistan and Afghanistan, ordered its supporters around the Arab world to go to Syria to help overthrow Assad. One of the groups that responded was al Qaeda in Iraq (or at least what remained of it), now run by an obscure self-proclaimed cleric known as Abu Bakr al Baghdadi. In addition to leading the remaining elements of al Qaeda in Iraq, al Baghdadi was part of the broader Sunni insurgency, which included some of Saddam's old military officers. The umbrella group had called itself the Islamic State of Iraq. Hunted down by U.S. Special Forces, they controlled no territory. To al Baghdadi, sending his fighters to Syria was effectively transferring them to a safer location.

President Assad began to paint his struggle to stay in power as a fight against fundamentalist, extremist terrorists. Secret police raids carried off hundreds of suspected activists, many of whom were executed without even the pretense of a trial. When in June 2011 word reached the U.S. Embassy of a mass grave, Ambassador Ford drove out of Damascus and went to the site, the international media in tow. In photographs from that day, Ford is seen with a cloth to his mouth to prevent him gagging from the stench of the rotting corpses.

By July, Assad, unable to quell the street rallies in Hama with police attacks, ordered the city surrounded by tanks and artillery. Ford feared the city would be leveled, as it had been thirty years earlier when the citizens had rebelled. Ford got in the car in Damascus and drove there, with one bodyguard along for the ride. He wore a baseball cap and flak jacket. He was literally in the "Arab street." And he was literally greeted with olive branches and flowers from the protestors as his car drove through the besieged city. Robert Ford was showing that America meant it when it said that it stood for human rights, for freedom, for democracy. Upon his return to Damascus, he continued sending his recommendations back to Washington, urging a tough line against the Syrian government. By August, President Obama was publicly calling for Assad to go. Assad was not pleased.

Assad's government complained that the American Ambassador was inciting violence. Assad's thugs tried to rough him up, pelting him with eggs and tomatoes. Then they assaulted the embassy, climbing over its walls and driving the U.S. diplomats and staff into a safe room. The police were nowhere to be seen. In response to Ford's urgent calls for security, officials of the Assad government said they would send over a unit to protect the compound. Hours went by. The handful of U.S. Marines with the Ambassador loaded their weapons, though Ford ordered them to hold fire. Eventually, the police arrived and the "protestors" left, but the embassy and its grounds were trashed.

In the U.S. and international media, Ford was widely portrayed as a hero, but some in the State Department thought he had brought on the embassy assault with his grandstanding. Others thought the next assault on the embassy would result in American casualties, maybe including Ford. He got the order: shut down the embassy. Ford returned home. He had been there only nine months. The Syrian Civil War had been raging for most of that time, but it was just beginning.

THE WRITING WAS ON THE WALL

Back in Washington, Ford commuted daily to his temporary office in the State Department. From there he lobbied the decision makers on the seventh floor of Foggy Bottom and in the White House on his proposals to aid the Syrian insurgents. He also spent time on the Hill briefing concerned congressmen.

Returning home was bittersweet. He loved his townhouse in Baltimore's Federal Hill neighborhood and enjoyed working closely with his stateside colleagues, but Robert Ford was a field man, not a bureaucrat. He felt most in his groove when he spoke Arabic more than English, when he could buy fresh vegetables and spices in the souk, as he had begun doing decades earlier as a Peace Corps volunteer in a poor Moroccan town. He was also used to a frenetic pace and to being near the action, or in it. After the U.S. invasion of Iraq, he had served in senior positions in the U.S. Embassy in the Green Zone for most of the eight years of U.S. occupation.

His time in Baghdad had given Ford unmatched insight into terrorism and Arab factionalism. Now, in Washington, he worried that what he had seen happen in Iraq could happen again in Syria. The extreme violence and discipline of the al Qaeda factions would allow them to take over the opposition movement. Already there were reports that a wealthy sheikh in Qatar was sending donations to the Nusra Front. It had emerged from the shadows, from the underground, from the camps in Afghanistan and Iraq. Disaffected Muslims were going to Turkey and then slipping across the border into Syria. The Turks were letting it happen. Maybe they were even helping, as there were reports that Turkey was quietly welcoming Syrian military officers who had defected from the Assad regime.

As the months went by and Assad ordered the army to use more extreme tactics against civilians, including artillery barrages and aerial bombing, thousands of Syrian military personnel deserted and began looking for groups that would fight the government. One such group called itself the Free Syrian Army. Although largely secular, it made common cause with the religious opposition elements who were members of the Muslim Brotherhood. They wanted Assad to join the list of overthrown Arab dictators in 2011.

As that year ended, chaos replaced dictatorial rule in Egypt, Libya, and Yemen. In Iraq, however, a multiethnic secular democracy seemed poised for success. U.S. troops and intelligence had almost entirely eliminated the threat from al Qaeda in Iraq and its odd bedfellows, the army officers of the Saddam Hussein regime. U.S. forces had also trained and equipped a large, new Iraqi Army. The new Iraqi leaders had refused to accept a residual U.S. force and the Obama administration had not tried hard to dissuade them. Thus, as the final days of 2011 ticked off, the last U.S. military units rolled across the border into Kuwait. For the first time in eight years, there were no U.S. combat troops in Iraq. What happened next was exactly what Western experts on Iraq had feared most.

The Iraqi President, a Shiite, turned on the leading Sunnis in the government and on the moderate Sunni militias that the U.S. had created to help defeat al Qaeda. In 2012, step by step, President Maliki and his supporters consolidated power by stripping Sunnis of any vestige of participation in government.[1] In Sunni-majority cities, resentment toward the Maliki government propagated, and protests were organized. In Washington, Ford, still technically the U.S. Ambassador to Syria, observed not only events in that country, but also those in Iraq. He knew the players there well. He knew that if the Shia kept up their retaliatory and exclusionary political tactics, the Sunni would eventually revert to armed resistance. Watching both Syria and Iraq unravel, Ford began to fear that the two separate issues could become one.

As 2012 wore on, the opposition to Assad in Syria not only remained resilient, it grew, fueled in part by the government's vicious
crackdown on the populace. Ford began finding support in Washington for his idea that the United States should aid the Syrian opposition
to make good Obama's call for Assad to go. The CIA developed a plan
for action. Ford and others in the State Department persuaded Secretary of State Hillary Clinton to support it. At the Pentagon, Secretary
of Defense Leon Panetta was on board too. The White House staff
conducted a series of interagency meetings to review the current options, but it seemed to be a slow roll. Somewhere in the White House
there was resistance. The meetings continued, but there was no action.

By the end of 2012, Ford saw clearly what would result from
failing to act. Assad would not be thrown out, but neither would he
quickly regain control of the rebellious cities and regions. The opposition was too broad and deep. Without U.S. support, the moderate Islamic and secular opposition could not control the rebellion.
The Nusra Front or some variant would step into the leadership role.
Thousands of Muslims were pouring into Syria to fight Assad. They
had two choices; they could fight for the moderate Free Syrian Army,
or they could fight for al Qaeda. Ford believed that if al Qaeda was
well-funded and equipped and the Free Syrian Army was not, the
volunteers would gravitate toward the terrorist group.

Worse than that, however, was the merging of the Syria and Iraq
situations. Elements of al Qaeda in Iraq had already moved into
Syria and were fighting there, gaining recruits, testing combat and
terrorist tactics. A strong al Qaeda–like force in Syria could, in turn,
strengthen what remained of al Qaeda in Iraq. Moreover, by that time
it was clear that Iraq was becoming fertile ground for radicalization of
the Sunni population. To them, the usurpation and crackdown by the
Shia in Baghdad didn't seem likely to stop.

Seeing the potential for the two conflicts merging, Ford had his

Cassandra moment. He issued a formal warning in writing to the Secretary of State, with copies for the White House. If the U.S. did not act quickly and decisively to train and equip the Syrian opposition, an al Qaeda–like group could seize and hold large swaths of territory, even major cities, on both sides of the Syria-Iraq border. The U.S. had invaded Afghanistan to eliminate an al Qaeda sanctuary. It had bombed al Qaeda's affiliates in Yemen and Somalia to prevent terrorists from acquiring territory. The lesson of 9/11 had been never to let a terrorist group establish a refuge from which it could launch attacks. Unless the U.S. moved quickly, Ford warned, that was exactly what was about to happen. Just to rub salt in the wound, the sanctuary would include parts of Iraq, where the U.S. had just spent a trillion dollars and suffered tens of thousands of American combat casualties.

John Kerry was confirmed as the new Secretary of State in January 2013. He seemed to get it, Ford thought, but the White House opposition to a major U.S. program in Syria had hardened. Administration officials were concerned that arms sent to the Free Syrian Army would end up going to al Qaeda or other extremist groups. They worried that the U.S. would be training people who were pretending to be part of the Free Syrian Army, only to later return to al Qaeda. A comprehensive vetting process would be needed before anyone got training. White House staff also cast doubt on the actual military nature of the Free Syrian Army, questioning the combat effectiveness of an opposition that included "former teachers and pharmacists." The White House wanted anyone trained by the U.S. to sign a pledge that they would fight only terrorists and would not try to conduct a regime change in Damascus. Ford was appalled. He thought there were no such people. He was right.

Throughout 2013, Ford was asked by congressmen and senators to explain what was going on in Syria. He dutifully explained the President's policy to skeptical members of Congress and the media, but what he saw happening in Syria and Iraq further convinced him that his grim scenario was coming true. The leader of al Qaeda in Iraq, al Baghdadi, unilaterally announced the merger of his organization with the Syrian al Qaeda affiliate, the Nusra Front. The new organization's name in Arabic was quickly translated by Western media into "the Islamic State of Iraq and Syria," or ISIS. The leaders of al Nusra refused to go along, so al Baghdadi broke with the al Qaeda leadership and pulled his troops out of al Nusra. Most of the foreign fighters came with him. Suddenly there was a single terrorist group fighting on both sides of the Iraq-Syria border, and they were winning. In July 2013, they had mounted attacks within a few miles of the Iraqi capital of Baghdad, taking over the infamous Abu Ghraib prison and freeing over five hundred detainees. For Ford, the writing was on the wall. He knew he had been right, even though he had wished otherwise.

The zenith of White House reluctance to get involved in Syria came in September when the President backed down at the last minute from previously planned airstrikes against Assad's chemical weapons. President Obama had previously said that the Syrian government's use of chemical weapons against its own people was a "red line" that would result in U.S. military action. But with undeniable proof that Assad had done just that and with U.S. aircraft poised to attack, the President and his chief of staff, Denis McDonough, decided to cancel the strikes. While the President's later diplomatic efforts to remove Syria's chemical weapons were successful, the bombing, Ford knew, would have done more than just destroy the chemical weapons. It would have greatly strengthened the morale of the opposition and the influence of the U.S. with the remnants of the Free Syria Army. The decision to cancel the attacks had the opposite effect, and President Obama immediately came under an immense amount of criticism for his change of plans.

Hosni Mubarak, the long-serving president-cum-dictator of Egypt, had been sacked by popular uprising, as had his Muslim Brotherhood successor, Mohamed Morsi. Secretary Kerry asked Ford to go to Cairo as U.S. Ambassador, for Egypt's Arab Spring had left the United States with a diplomatic mess. For Robert Ford, Cairo was another choice ambassadorial post, but sitting in his State Department office pondering the decision, he fumed. There was no doubt that he felt a commitment to his government, and that he truly felt the Foreign Service was an honorable and noble way to serve his country. Egypt would be his third ambassadorship, after Syria and Algeria. Instead of taking the position, however, he resigned from the U.S. Foreign Service. "I couldn't keep supporting a Syria policy I did not believe in," Ford told us. Moreover, he thought that by staying in Washington he could advocate a better policy toward Syria and perhaps the region. Ford accepted an offer to be a senior fellow at one of the oldest foreign policy think tanks in Washington, the Middle East Institute.[2]

Throughout 2013, the Ford scenario continued to play out as he had predicted. The Nusra Front gained ground in Syria, supported by foreign fighters, arms, and funds from wealthy supporters in the Arab world, as well as the governments of Qatar and Turkey. More worrying, ISIS was gaining more ground than Nusra and the even weaker Free Syrian Army. Assad's army was cracking under all of the pressure, but remained propped up by Iranian special forces. In Iraq, the Shia government was in full attack mode against Sunnis, using force to quell demonstrations and ejecting Sunni officials from the government, security services, and army. Some Sunnis, including former Saddam-era military officers, began reluctantly opening talks with ISIS. Eventually, it was those Saddam military men who were responsible for taking many of the new foreign fighters in Syria and turning them into an effective combat unit for ISIS. There were enough new recruits pouring into Syria that ISIS began sending some of them into Iraq.

As 2014 opened, the Ford scenario was not a theory anymore. It was front-page news. ISIS took Raqqa on January 2, proclaiming the Syrian city the capital of the Islamic State. The next day, they attacked in Iraq, assaulting and seizing Fallujah, a Sunni city of over three hundred thousand people. Fallujah had been the scene of the bloodiest battle in the second U.S.-Iraq war, a place where hundreds of Americans were killed and wounded. Now, the city that Americans had died to liberate from al Qaeda was completely controlled by the successor of that organization, ISIS.

In Washington there was a sense of panic. The CIA had not seen these attacks coming. The Pentagon was amazed that the Iraqi Army it had trained and equipped had crumbled at Fallujah. At State, the regional experts were less surprised. They still remembered Ford's prediction. Unlike the 1990 Iraq-Kuwait showdown, this was a civil war, and understanding the culture and region as Ford did was key.

ISIS advanced quickly, capturing towns and oil wells in western Iraq. The Iraqi Army regrouped, but could not manage to stage a counterattack on Fallujah. Then there was another surprise ISIS attack in early June, this time on Mosul, Iraq's second-largest city with over a million people. Again, the Iraqi Army crumbled. In Raqqa, ISIS had already begun setting up government agencies to administer and govern the cities and the oil wells under its control. By July, al Baghdadi appeared in the main mosque in Mosul and proclaimed himself Caliph of the new Islamic State. No longer limiting his ambitions to Iraq and Syria, he saw the Islamic State as eventually extending across the Muslim world.

ISIS's army moved on, threatening the Kurdish city of Irbil, home to a U.S. Consulate and military training mission. Reluctantly, President Obama ordered air strikes to protect the Kurds. ISIS responded by releasing gruesome videos of American prisoners being beheaded.

Back in 2013, the President had referred to ISIS as a "junior varsity" team and didn't seem worried. By September 2014, Obama vowed to "degrade and destroy ISIS." Later that month, he ordered a sustained bombing campaign against it in both Iraq and Syria.

Despite the President's pledge, the group hung on to its major cities and continued successfully to combat the Syrian and Iraqi armies. Hundreds of U.S. military planners and advisors returned to Iraq. Plans for the Iraqi Army to evict ISIS from Fallujah and Mosul in the spring of 2015 fizzled. ISIS took another Iraqi city, this time Ramadi, with a population of over two hundred thousand. The U.S. did eventually try to assemble a vetted and armed Syrian resistance group, but it was wiped out by an al Qaeda affiliate group's surprise attack before it could even begin to operate.

Under Osama bin Laden, al Qaeda never truly controlled a single sizable town. Under al Baghdadi, ISIS has numerous towns and cities, spread across what used to be the Syria-Iraq border. That demarcation was now essentially gone, made meaningless by the creation of the new nation, the Islamic State. It was equipped with U.S.-made artillery and armored vehicles stolen in Iraq, funded by oil sales from wells under its control, and staffed by thousands of fighters from throughout the Muslim world. That "state" was governed by (not just influenced by) a terrorist organization and had aggressive plans to expand its revolution elsewhere.

After 9/11, we said terrorists would never again be allowed a sanctuary, but, again, they had one. "Lone wolf" terrorists in the U.S. and Europe now conduct violence in the name of ISIS. Groups in Central Asia and Africa have affiliated themselves with the new Islamic State. ISIS has become more than just a nation-state in the eyes of these radicalized lone wolves. It has become an idea and a vision of the future that they deem appealing and worth fighting for.

THE SLIPPERY SLOPE OF INTERVENTION

Sitting by the shore of the Arab Gulf on a warm night early in 2016, we asked Robert Ford why his warning and prediction had become another tragic Cassandra Event. Why had people in power not listened? "Oh, they listened," he said. "And many of them got it, including Secretary Clinton and then Secretary Kerry."[3] They may not have believed that a single new terrorist state would arise spanning Syria and Iraq, but they saw that in the absence of a strong American hand supporting the Syrian rebels, the vacuum would be filled by others inimical to U.S. interests and values.

Sitting with this preeminent American specialist on the Arab world, it's easy to see him as an academic. He is soft-spoken and precise in his words, not a bomb thrower, fearmonger, or hysteric. Those who worked with him for years told us that he is likable and quickly engenders respect. Far from actually being an academic, however, he was a fearless practitioner, not afraid of getting his shoes dusty or maybe his car shot up in the world beyond the embassy's walls. His Arabic, they said, was near native, and his understanding of contemporary Arab culture and politics was unsurpassed in the U.S. government.

Why then, we asked, had no action been taken in time to stop his prediction from coming true? Robert Ford believed that the reluctance of White House staffers he saw in meetings, and their slow-rolling of the interagency process, was a reflection of their boss's feelings on the matter. The President, Ford realized, thought that the U.S.-Iraq war had been a colossal mistake and seriously doubted the ability of the U.S. to create desirable outcomes in the Middle East, especially through the use of military force. Aiding the Syrian opposition in any significant way was a slippery slope to another U.S. military intervention, the President thought. Obama had been persuaded years earlier to get involved in Libya, ordering extensive U.S. airstrikes against the Qaddafi regime. That resulted in a quagmire, continuing violence, growing terrorist groups, and a failed state. He did not want to get involved militarily in the Middle East again, especially since he had finally pulled U.S. troops out of Iraq and was pushing hard politically to get them out of Afghanistan too.

Ford thought Obama's fear of a slippery slope was unjustified. He contrasted Obama's fear with the kind of carrot-and-stick diplomacy that President Clinton had conducted with the help of Ambassador Richard Holbrooke in the Balkans during the 1990s. There, the U.S. had employed airstrikes and sent in the army. "I never called for the use of U.S. forces in Syria. If we had acted in time to arm the opposition, that would not have been needed," Ford explained. He believes that in 2011 and 2012 the moderate opposition was filled with Syrian Army defectors, military men, not just schoolteachers and pharmacists. Had we acted early to arm them, they could have attracted an even larger group of fighters, he says, but even so, the White House staff (and likely the President) doubted that the moderate opposition would have any significant impact on the fighting.

This Cassandra Event is a clear case of a decision maker failing to heed the expert, rejecting a Cassandra's analysis and recommendations out of hand. The decision maker, in this case President Obama, saw the immediate situation in a larger context and believed the benefits of acting were outweighed by the risks and perhaps the opportunity costs. Fareed Zakaria, who interviewed the President about the situation in Syria and Iraq, told us that the impression he got from the body language and comments on the margins of the interview indicated that Obama thought that regional experts had been wrong before about the Middle East and were wrong again, and that there was nothing he could do to dissuade the region's factions from killing each other. Moreover, a critical component of his campaign platforms in 2008 and 2012 was the belief that America needed to end the wars in the Middle East. How, after that commitment to the American people, could he allow himself to get dragged into another war?

Ford and his colleagues had warned about a terrorist-controlled zone being created spanning the Iraqi-Syrian border, but no one else had told the President. Mike Morrell, a career CIA analyst who rose to the top ranks of the Agency, admitted to us that the CIA didn't assess it as a possibility. Morrell told us that the CIA underestimated ISIS and overestimated the Iraqi Army. Had the CIA, regional leaders, or even op-ed pundits joined in Ford's forecast, Obama might have reacted differently.

Ford believes he became a Cassandra partly because the decision maker did not understand the underlying dynamics of the region and held a distaste for dealing with its wars. The President was also influenced by his political commitments, what we might think of as an ideological filter. Such a filter is similar to what we saw in our Hurricane Katrina case study, when the Bush administration's single focus on terrorism reduced the resources and effort that could have been put toward preventing the destruction of New Orleans. Moreover, nothing like this had ever happened before, making this case yet another example where Initial Occurrence Syndrome played a role in the minds of decision makers. Officials and analysts saw the possibility of a protracted Syrian civil war, yes, but most didn't see it spawning an extremist nation-state. Ford had seen it, though, and he had warned about the problem of "foreign fighters" returning to Europe or other home countries after learning how to fight in Syria and Iraq. What he did not see, Ford told us, was the size of the refugee flow or its political effect in Europe. Nor, he acknowledged, did he imagine how large the Iranian role would become, or how Russia would prevent Assad's fall by direct military intervention.

By October 2015, the validity of the predictions Ford made was obvious. Then a private citizen and a scholar at the Middle East Institute, Ford was in a meeting at the United Nations when his mobile phone vibrated. The caller ID was blocked, but he took the call. It was the White House. The voice at the other end wanted to know if he could come to see President Obama. Ford hoped that the President had changed his mind, but in their two-hour meeting, joined by former U.S. Ambassador Ryan Crocker, Obama was defensive, talking more than listening, justifying his policy. Ford summarized the President's position to us as "it's too bad what happened to Syria, but there was nothing America could have done about it." Although thousands of Syrians had been killed by barrel bombs and chlorine, the President took solace in the fact that it could have been worse had he not arranged for the withdrawal of nerve gas from Syria. Far from evincing self-doubt, Obama wanted to explain why he still thought he had been right all along about Syria. His biggest foreign policy mistake, Obama had told reporters, was his handling of Libya after the fall of Qaddafi. Libya, not Syria.

A year later, in October 2016, when we sat down with Ford again, he noted that Obama and his close associates always had objections to various courses of action presented to the President about what to do in Syria. They seemed to want only a perfect, flawless option. Never, Ford noted, had they given equal critical analysis of the option that they had implicitly chosen, essentially the path of inaction. "Doing nothing is an option. It too has consequences," he told us. In this case, doing nothing had created results far worse than anything that Ford could have imagined as the negative consequences of U.S. action: millions of refugees and internally displaced people, cities in rubble,

Europe dealing with a refugee crisis which disrupted its politics and fractured its unity, tens of thousands of civilians dead, Iran and Russia ascendant in the Middle East.

Daesh, ISIS, was finally being ejected from towns and some cities in Iraq and Syria by the end of 2016. Perhaps, we suggested, the crisis which we might have averted will soon end? "No," Ford replied, "we will not see a unitary government in control of all of Syria again in our lifetime. Daesh will not go away when it loses its cities. The funds to rebuild the cities and to allow the refugees to return will not materialize. The refugee camps and the ruins of parts of Syria and Iraq will be a poisoned breeding ground for the next wave of terrorists." Behind his professional face, we could see the sadness, the pain, the empathy for the Syrian people, the dead and the displaced, made all the worse for his belief that it did not have to be this bad.

CHAPTER 5

The Seismologist: Fukushima Nuclear Disaster

Remember the calamity of the great tsunamis. Do not build your homes below this point!

—CENTURIES-OLD STONE TABLET, ANEYOSHI VILLAGE, IWATE PREFECTURE, JAPAN

Waves and castles, high waves and castles destroyed, the images kept going through Yukinobu Okamura's mind. Images of destruction from the distant past haunted him, but he feared for the future, the very near future. It was June 2009.

Dr. Okamura listened carefully as a panel reported on the readiness of Japan's Fukushima Daiichi nuclear power plant to withstand a severe natural disaster. Okamura was a noted seismologist whose opinions and warnings the panel ought to heed. At least he hoped they would. He had something new and important to share.

Japan's Nuclear and Industrial Safety Agency (NISA) was holding meetings to discuss the particular safety needs of each of the country's seventeen nuclear power plants. The nuclear reactor campus in the Fukushima Province lay outside of the small provincial capital city of the same name, a seaside city of three hundred thousand people about 125 miles north of Tokyo, on Japan's east coast. Fukushima Daiichi, the six-reactor nuclear power plant complex, was owned and operated by Tokyo Electric Power Company (TEPCO). For TEPCO, safety at Fukushima meant focusing on earthquakes.

Earthquakes are a very real and significant risk to nuclear power plants, warranting special measures. Memories of the Chūetsu offshore earthquake of 2007 were still fresh. The magnitude 6.6 quake had sent tremors from Niigata, in western Japan, all the way to Tokyo. It caused a fire to break out in a transformer building at the world's largest nuclear complex, Kashiwazaki-Kariwa, another nuclear plant run by TEPCO. Radioactive gases leaked. Water from a pool of spent nuclear fuel entered the sea. The plant was effectively shut down for two years. TEPCO did not want that to happen again. Neither did the government.

With this recent event in the forefront of everyone's mind, the Fukushima panel's focus on earthquake geology was understandable. Its members sought to assess whether and to what extent Fukushima faced similar vulnerabilities. Before presenting its findings at the larger meeting where Okamura voiced his concerns, the panel had met twenty-two times. Tsunamis were never on the agenda. NISA, which predetermined most of each panel's considerations, did not see tsunamis as likely enough in the Fukushima region to warrant consideration. "An operator of nuclear plants needs to take a precaution even against an extremely rare natural disaster," Okamura later asserted. "Even if there was only a slight chance," he added, TEPCO "should have taken action."[1] But the seven-member panel did not include a tsunami expert.

In creating safety guidelines for Fukushima, the panel used data from the largest earthquake recorded in the area, a major one from 1938, one that measured 7.9 on the Richter scale. That earthquake had caused a small tsunami and, at Fukushima, the reactors were close to the sea. Therefore, TEPCO thought that a seawall was necessary at Fukushima, in case that kind of tsunami happened again. They built a wall about nineteen feet above ground. It would have stopped the 1938 tsunami from flooding the reactor complex.

Dr. Okamura is a short man with a quiet voice, and it took considerable effort to convince him to share any of his story with us. He is a well-respected expert, director of Japan's Active Fault and Earthquake Research Center, but not a famous scientist. Okamura had been what many Japanese call a salary man throughout his respectable career, doing his work well, without any splash or complaint, and slowly rising through the ranks. But on this day in June 2009, this quiet little man had something big to say, and since then Okamura has regretted not saying it louder. Considering the threat of a significant tsunami in the region, he later reflected, "The truth is, I felt it was high time"[2] to speak up.

When Okamura addressed the panel, he began by declaring that the 1938 quake was simply not big enough to serve as a basis for the Fukushima guidelines. "I think you know this," he said, "but I would like to confirm that this [1938 earthquake struck at] the site of the Jōgan Earthquake, or Jōgan tsunami, if you like." Jōgan, this AD 869 earthquake, Okamura continued, was known to be "overwhelmingly greater" than the 1938 quake. "I would like to know," Okamura continued, attempting to remain respectful and composed in what would prove to be a critical moment, "why you didn't mention this at all?"

TEPCO representative Isao Nishimura responded that the Jōgan quake "did not cause much damage." But Okamura insisted that the damage had been severe; it had been enough to wipe out a castle. "So I think," he hastened to add, "there is no evidence for you on which to base the statement that there wasn't much damage."[3] Okamura said that he was particularly concerned about the Jōgan tsunami, which had reached far inland. Another like it could certainly threaten the Fukushima region. "That information is available to us," Okamura told Nishimura. "But you are not mentioning it at all. That is what I can't understand."[4]

Like Charlie Allen and Ivor van Heerden but perhaps even more explicitly, Okamura was asking the authorities, "Why are you ignoring incontrovertible data?" The data, more than any other factor, drove Dr. Okamura's concerns and his warnings. (Clearly, a theme is emerging.)

That day, however, both the government and TEPCO insisted that earthquakes were the threat, the only threat, upon which to focus. Because this series of meetings had been designed with particular aims in mind and the particular aim of discussing Fukushima that day was to focus on the earthquake threat alone, tsunamis had no place in the conversation. No one suggested an alternate time to address the threat of tsunamis. Besides, a TEPCO executive noted dismissively, the 869 earthquake was simply "historical." Why attempt to extrapolate information from a retold story from so long ago, a story not backed by today's sophisticated measurements and data, when the much more recent 1938 earthquake would do the job just fine?

The meeting moved on. TEPCO officials said they would try to learn more. When the next meeting came around, Okamura was disappointed to see that the safety report for Fukushima was approved and that the 869 earthquake had been downplayed as not very destructive. When they allowed him to the microphone, again he pressed the issue, but again his concerns were dismissed. "I thought TEPCO understood [the risk] as they said they were going to examine it," he pleaded to the panel, genuinely puzzled as to the lack of investigation. The "historical" earthquake and tsunami of 869 would not be the model for the updated Fukushima Daiichi safety guidelines.

TRUTH IN THE SAND

The same thing that motivated so many other Cassandras whose stories we researched also motivated Okamura: hard facts and new data that others chose to discount or ignore. His institute, the Active Fault and Earthquake Research Center, was able to generate new data on the 869 earthquake, data that no one else had examined. Though his warnings were ignored at those fateful meetings, this Cassandra used the geologic past to see the future, and he had on his side the science to which he had devoted his career.

The events of the 869 earthquake and tsunami, historically called Jōgan Jishin in Japan, were recounted in the Japanese history text *Nihon Sandai Jitsuroku* [True History of Three Reigns of Japan], compiled in 901. "People shouted and cried, lay down and could not stand up," the book reports, following a "large earthquake" in Mutsu Province, during which there was "some strange light in the sky." People died in landslides or were crushed by the collapse of their homes. Buildings, gates, and walls were destroyed. "Then the sea began roaring like a big thunderstorm. The sea surface suddenly rose up, and the huge waves attacked the land."

The description from twelve centuries ago is still chilling: The waves "raged like nightmares, and immediately reached the city center. The waves spread thousands of yards from the beach, and we could not see how large the devastated area was. The fields and roads sank into the sea." Those who did not escape, about a thousand people, drowned.

Okamura's critics took a contemporary perspective and pooh-poohed this account to be only a story, a story that feels, because it happened so long ago, like a legend. And, they seemed to say, we cannot base our science on legend. It is not good science, not usable data.

This reaction is what we will call Scientific Reticence, a reluctance to make a judgment in the absence of perfect and complete data.

In this case, it was married to availability bias, that all-too-frequent human cognitive error whereby an individual is overinfluenced by information already known, as in the case of Dr. van Heerden and Hurricane Katrina. The Fukushima panel did know of the 869 tsunami, but it was so distant that they had no emotional connection to it, so they didn't incorporate the destructiveness and reach of Jōgan Jishin into their calculus. The availability bias is partly why the NISA panel was quick to dismiss Jōgan Jishin as unworthy of attention. However, as we will discuss, more short-sighted reasons also fueled what may have been willful ignorance on the part of the regulators.

The scientists at Okamura's institute knew the 869 tsunami was much more than a myth. The causal earthquake, and many others, had occurred at the convergent boundary between the Okhotsk Plate and the subducting Pacific Plateau, a location, tectonically speaking, with a lot going on. This site is prone to large earthquakes, many of which have triggered tsunamis. And the Active Fault and Earthquake Research Center had recently carried out detailed surveys of sand deposits left behind by Jōgan Jishin.

Before the surveys, the extent of the damage from the Jōgan Jishin quake was not clear. Most of what was known about the quake and the tsunamis that followed came from historical records, not geological ones. Okamura's center was among the first to analyze the deposits left behind by tsunami waves and to then map the flood zone. From their research, Okamura and his colleagues concluded that the tsunami had led to extensive flooding of the Sendai Plain, at least 2.5 miles inland. The area inundated by Jōgan Jishin, according to the dated deposits, would later prove to match very closely the area flooded by the tsunami that struck Fukushima Daiichi in 2011.

Okamura and his group had modeled what would happen if a magnitude 8.4 earthquake were to strike offshore. Their model showed that waves higher than twenty feet would hit the Fukushima coast. By

2007, the research center had already shared much of this data in a study published by the Geological Survey of Japan. In 2009, another study found the possibility of even larger waves: up to thirty feet.

Doing such innovative work was exciting for Okamura, and he was eager to share his findings, because of what they revealed about the Jōgan Jishin tsunamis and earlier ones but even more because of what they could tell us about future threats. As Okamura told us, his "purpose was to warn residents" along the coastlines. For Okamura, the truth was now visible, clearly written in the sand deposits spread up and down the hills of Japan.[5] However, just as the scientists of the Louisiana coast found out, being in the vanguard of new analysis may also mean that your audience isn't ready to hear what you have to say.

That truth buzzed with the urgency of an alarm clock. As seasoned seismologists like Okamura well know, tectonic plates seem to follow regular schedules to unleash their earthquakes, and the deposits the Institute identified suggested that such earthquakes tended to occur about every 800 to 1,100 years. He did the math: 2011− 869 = 1,142 years. The numbers made one thing clear: the likelihood was high that a similarly large earthquake and tsunami would soon strike. It was already overdue.

It came as a surprise to most Japanese. At 2:46 p.m. on March 11, 2011, a magnitude 9.0 earthquake shook Japan. What would later be known as the Great East Japan Earthquake (or Tohoku Earthquake) struck eighty miles east of Sendai in Honshu, Japan, about eighteen miles below the surface of the water. The quake caused three to five minutes of vigorous shaking, and it released a series of tsunami waves from the depths of the Pacific Ocean. These tsunamis, which raced at 435 miles per hour, exploded onto Japan's shore with a force not experienced for at least a thousand years.

The walls of water and accumulated debris roared through coastal towns, schools, shopping malls, resorts, and stadiums, making their way six miles inland, inundating hundreds of square miles of the coastal region. Residents of Sendai had only eight to ten minutes' warning before the tsunami hit. Tragically and ironically, more than a hundred evacuation sites, built to withstand the force of the 1938 tsunami, to which many citizens fled seeking safety, were enveloped by the waves. Many people at tsunami evacuation sites were crushed on impact or washed away to death by drowning.

This was the globe's fourth strongest earthquake since modern civilization began recording measurements (around 1900), and the most powerful to hit Japan in recorded history. People reported feeling its tremors from over 1,000 miles away. The quake moved the entire island of Honshu about eight feet to the east, and 250 miles of the coastline experienced a vertical drop of two feet. That drop allowed the tsunami to have an even greater impact, hastening its movement farther inland. Hundreds of thousands of buildings collapsed or were damaged. In Sendai, the inundation destroyed homes within seconds. More than nineteen thousand people were killed.

But it was no great surprise to Dr. Okamura, who was in Tsukuba, a city in Japan's Ibaraki Prefecture (a little over a hundred miles from Fukushima) when the earthquake hit. The man who had urged further consideration of tsunami risk, who "could not understand" why so many had ignored his warnings, was powerless to stop the terror he had foreseen. He was more horrified than ever that he had failed to convince anyone to listen, because he knew what would happen next.

TICKLING THE DRAGON'S TAIL

During the Manhattan Project, America's nuclear scientists were putting atomic theory into practice for the first time. Working with their bare hands in the New Mexico desert heat, and wearing nothing

more than dungarees and T-shirts, the scientists would cover plutonium spheres, known as pits in the industry, with various materials to
discover the optimum ways to reflect the radioactivity from the plutonium back into itself. Reflecting the radiation back into the pit is the
first step to creating a nuclear chain reaction and an eventual atomic
explosion.

The scientists would place reflectors, consisting of metal bricks
and thin beryllium hemispheres, around the pit with their hands and
use flathead screwdrivers to adjust the distance between the reflectors
and the plutonium. The emitted neutrons were reflected back into
the pits, and the scientists could then measure the crackling of the
chain reaction with their Geiger counters, thus developing a better
understanding of when the pit would go critical and exactly what was
needed for a nuclear explosion.

The researchers called this exercise tickling the dragon's tail, for
it was extraordinarily dangerous.[6] Within a period of one year, the
dragon awoke twice and each time released a massive deadly flash of
neutron radiation, a "supercritical" release of energy. In the first instance, a scientist dropped a brick on the pit and was dead three ago-
nizing weeks later. In the second, the scientist's screwdriver slipped,
and a neutron-reflecting shield clapped down onto the pit. He was
dead in nine days. This particular pit came to be called the Demon
Core.[7]

In each instance, an ominous blue light flashed from the Demon
Core as the massive surge of neutron and gamma radiation ionized
the air. In total, both incidents irradiated ten people who later died
(some almost immediately and some many years later from radiation-
derived cancers).

Nuclear reactors work on a similar principle but with the element
uranium instead of plutonium. They tease the radioactive material
to release waves of energy but do it in a much more controlled and

safer manner than those young scientists using bare hands and screw-drivers. Most modern reactors surround fuel rods of uranium with control rods made of various metals that absorb the neutron emissions to the exact degree needed. The entire assembly is kept under a continuously recirculating pool of water to constantly transfer heat away from the rods. The ability to move the control rods and to circulate the water in the cooling pool is necessary to prevent the rods from overheating and melting the uranium of which they are made. That phenomenon is notoriously known as a meltdown.

WHAT HAPPENS WHEN NUCLEAR POWER PLANTS HAVE NO POWER?

The bombs developed by the bright minds in that New Mexico desert brought radioactive death to Japan in 1945 and created a cultural memory still strong among the Japanese people. When Prime Minister Naoto Kan called the 2011 earthquake, tsunami, and ensuing nuclear disaster the "toughest and most difficult crisis for Japan" since the end of World War II, he was evoking that memory. Sixty-five years after the nuclear attacks, those atomic bombs could not have been far from the prime minister's mind, or the public's. What Okamura had predicted had happened: another nuclear disaster for the Japanese people. This one, however, was self-made.

The reactor units at the Fukushima Daiichi power plant responded to the earthquake as they were meant to do. Seismic sensors recorded the earthquake and instantly caused neutron-absorbing control rods to spring up from within the reactor cores to stop further nuclear fission. The safety systems performed precisely as expected. At sea, however, the earthquake had created a tsunami. It moved west, toward the Japanese shore, toward Fukushima. When it hit the beach, it continued inland and over the nuclear plant's nineteen-foot concrete seawall, in places causing the wall to explode like a pane of glass struck by a truck. One wave, estimated at nearly fifty feet high and traveling faster than a landing jetliner, flooded the power plant an hour after the earthquake hit. Eight minutes later, more waves arrived, inundating the site. The millions of tons of seawater injected by the tsunami engulfed the plant's low-lying buildings.

Waves flowed quickly across the road and into the six buildings containing the plant's turbines and the six nuclear reactors. Before the water hit, the electricity coming into the plant from the electrical grid had been cut off. The earthquake had toppled the electrical towers, snapping lines. Conventionally powered backup generators (emergency diesel generators, or EDGs) at the Fukushima plant automatically kicked on, keeping power to the reactors for the sake of continuing to circulate water by the fuel rods to give them time to cool. Then the waves flooded all but one of the EDGs. There was almost no electric power at the plant, no way to control the reactors, no power for the pumps designed to clear out flood water from the buildings. Inside most of the six big reactor buildings, it was dark except for wan beams from battery-powered emergency lights.

The control rods had moved into place after the initial quake and before the power failure, successfully ending the nuclear fission in the reactors. But cooling fluids still needed to be pumped; otherwise the uranium fuel continues to radiate so much "decay heat" that the water in the reactor boils. Boiling leads to steam, which can build up massive amounts of pressure in pipes and in the important containment vessel around the core. That pressure can, in turn, eventually cause the containment vessel to break or even explode, emitting dangerous radioactive gases and particles into the campus buildings and far beyond, out into the neighboring towns, across the beach into the seas, and up into the drifting winds of the atmosphere.

Nuclear power plants are highly engineered cathedrals to humanity's supposed mastery of a power like that of the sun. Surfaces in the plants are as white as newly fallen snow, or they're stainless steel and shined daily. Plant management teams wear white doctors' coats and are highly trained, well-paid professionals. They follow precise protocols, with no deviation, every day.

What the Fukushima staff saw after the tsunami horrified them. Hours went by as the TEPCO workers tried to find ways to generate more electricity to turn on the pumps needed to cool the reactors. To their great relief, a generator truck finally arrived, but it turned out to have plugs that were not compatible with the plant's and was thus useless. Reactor 1's backup condenser failed, and pressure within the containment vessel rose, eventually going above safety limits. Plant managers knew, though their trusted protocols said otherwise, that they would need to vent the radioactive steam by creating an intentional leak; otherwise the massive steel and concrete containment vessel was at risk of exploding.

Conditions continued to worsen. Hydrogen leaking from the core of Reactor 1 mixed with oxygen, causing an explosion at 3:36 p.m. that Saturday afternoon. Flashing from cameras on helicopters offshore to television screens in Tokyo and around the world, the image of the building erupting looked like something from a Hollywood disaster movie. As the public watched in disbelief, radioactive particles were being pushed up into the winds, blowing toward Tokyo.

Over the next few days, troubles continued as TEPCO technicians intentionally vented steam from the reactors and tried to cool them by using fire hoses to shoot seawater directly into the overheating nuclear cores. Nonetheless, on Monday, a second explosion blew up the Reactor 3 building. The anxiety already gripping TEPCO executives, the Tokyo government, and many in the general Japanese population grew to general panic.

Soon after, problems were reported with the used-uranium storage pool at Reactor 4. Used uranium rods still give off heat. They are stored in tanks, into which cool water is pumped on a regular basis. In Reactor 4's storage tanks, the pump had no power. The fluid around the spent rods was heating up, threatening to boil away. If it did, the uranium rods would begin melting, catch fire, and release a tremendous amount of radioactivity into the air. On Tuesday morning, this is exactly what happened. The zirconium coating on the spent fuel rods in the Reactor 4 spent-fuel pool did, in fact, catch fire. Then another televised hydrogen explosion occurred, blowing apart a containment building, causing the greatest release of radioactivity in the entire cascade of calamities.

The Tokyo government mandated a twenty-kilometer (about twelve-mile) evacuation zone surrounding Fukushima Daiichi. The United States recommended that its citizens retreat even farther, to at least fifty miles away. Rumors spread, including one that a "radioactive cloud" would soon hit the West Coast of the United States.

Meanwhile, plant workers desperately continued to try to cool the reactors. On Wednesday, military helicopters attempted to dump seawater on the Reactor 3 spent-fuel pool but had to turn back because dangerously high levels of radiation put the pilots and crews at an unacceptable level of risk. That night, military fire trucks arrived and began spraying water at high pressure at the core. Japanese and people around the world were spellbound by the repeated images of explosions at the nuclear campus. They knew already that it was the worst nuclear disaster since the meltdown at Chernobyl in 1986. Perhaps it was worse.

Over the weeks that followed, radioactive water was found to be leaking into the sea. Workers scrambled to stop it. Around the world, fear and doubt regarding nuclear power grew. For weeks, hundreds at TEPCO worked to reinstate regular cooling at the plant. Over seven hundred "nonessential" workers were sent home four days after the earthquake struck. The fifty who stayed were lauded for their bravery. The media called them the Fukushima Fifty, and the number fifty stuck despite the fact that additional workers then arrived from all over the country, including firefighters and soldiers, to join the effort.

In a country whose citizens deeply value stoicism and determination, the workers quickly came to be regarded as heroes. The older workers courageously stayed, urging younger employees to go home, knowing they risked radiation poisoning. By March 23, more than a thousand people were working at the site. Prime Minister Naoto Kan commended the Fukushima Fifty for "making their best effort without even thinking twice about the danger." Two TEPCO employees died as a result of injuries following the earthquake, and six others received radiation doses above the regulatory lifetime limit.

Two years after the accident, the UN Scientific Committee on the Effects of Atomic Radiation (UNSCEAR) concluded a study that determined that the radiation exposure "did not cause any immediate health effects," though other studies contended that up to thirty-three thousand people will eventually die from cancers they contracted from the Fukushima radiation. The World Bank found the earthquake and tsunami to be the most expensive natural disaster of all time, with an economic cost estimated at $235 billion. Of the more than nineteen thousand people who perished in the disaster, most drowned or were battered to death by the flood.

In response to widespread public fear, Japan's prime minister ordered a "temporary" safety inspection and shutdown of all nuclear power plants. Plans to turn plants back on met with public resistance. In this country that once depended on nuclear power to meet 30 percent of its energy needs, no nuclear reactors operated at all for a full two years after the disaster. Today only one of Japan's fifty nuclear reactors continues to operate. Other countries took a cue from Japan's disaster, most notably Germany, which has pledged to phase out its nuclear program entirely by 2022.

Much of the region surrounding the Fukushima-Daiichi plant remains a wasteland of radioactive ghost towns. Cleanup and decontamination efforts around the plant are ongoing, though progress is painstakingly slow on a project so large. As of March 2016, about 171,000 tsunami refugees had yet to return to their homes, and many will never be able to do so.[8] The full decommissioning effort at the plant itself is expected to take up to four decades.

AN INSURANCE POLICY THAT PAYS OUT 2,000 TIMES THE COST

A Stanford study in 2013 concluded that any one of three key improvements could have mitigated or prevented all this from happening at Fukushima Daiichi: plant elevation, seawall height, or the relocation of the plant's backup generators. Higher seawalls would have prevented the tsunami's waters from spilling over into the plant, even situated where it was and at the elevation it was. Alternatively, higher plant elevation would have prevented the tsunami from damaging key components, such as the backup generators. Even just elevating and waterproofing the backup generators, or not installing them in the basement in the first place, would have made a major difference and could have potentially averted the crisis altogether. Okamura had suggested all three things in 2009.

In one of history's best proofs that "it doesn't pay to cut corners," TEPCO decided that it could save some construction cost if it lowered the land upon which it built the plant. In 1967, TEPCO sliced eighty feet from the natural seaside at the site to more easily transport equipment. Although this may have made the construction process easier and cheaper, it meant that the reactor buildings were not at the original height of the terrain, over a hundred feet above sea level, but rather at only thirty feet, dropping the nuclear reactors right into the tsunami's path. In 1967, TEPCO put short-term cost savings ahead of safety concerns, and their logic only seemed to get worse as the years went on.

Where used elsewhere, higher seawalls worked. At a plant in Onagawa, which was closer to the epicenter of the quake and hit by a tsunami of similar height, there was relatively little damage. Onagawa had a forty-six-foot seawall, which proved adequate to keep out a forty-two-foot tsunami. The nineteen feet above ground (and thirty-five-feet above sea level) seawalls at Fukushima Daiichi did not.

Perhaps most galling about this horrendous disaster is how small and relatively inexpensive the preventative measures to heed Okamura's warnings could have been. Precise studies are not available, but it likely would have cost TEPCO about $50 million to emplace an effective combination of better backup generators, waterproofing, and/or higher seawalls. Considering that the nuclear disaster alone will cost Japan at least $100 billion (not to mention the future cancers and the 1,600 who are estimated to have died from the stresses of the evacuation), ignoring Okamura and refusing to spend on safety resulted in damage over two thousand times more expensive. In other words, for a one-time expense of just 0.05 percent of the damage costs, TEPCO might have completely avoided the reactor disaster.

The Onagawa plant with the forty-six-foot seawalls has Yanosuke Hirai to thank for keeping it safe. Hirai, who received the Japanese Medal of Honor for public service in 1961, was an extremely responsible and safety-conscious civil engineer. But unlike Cassandra and Okamura, Hirai was heeded. He insisted upon holding his corner of the nuclear power industry to the highest standards possible, going above and beyond the industry's regulations. He saw safety as an essential part of his duty. His sense of responsibility and respect for the dangers of natural disasters saved many lives long after his own ended in 1986 at the age of eighty-four.

In 1968, when he joined a committee that was planning the construction of a plant for Tohoku Electric Power (of which Hirai had been vice president before his retirement), Hirai told the board that the plant should be built at an elevation of nearly fifty feet above sea level and be protected by a seawall forty-nine feet high. The original designs called for a seawall of only ten feet. Though others claimed thirty-nine feet would be sufficient, Hirai stood his ground. The final height chosen was 46 feet.

When the Great East Japan Earthquake hit Onagawa, the seawall remained intact and the tsunami stayed out. Hirai, like Okamura, made his choices with the "historical" Jōgan Jishin earthquake and tsunami of 869 in mind. As a child, he had visited a shrine that had been hit hard by that tsunami, and the memory stayed with him. His visit to that shrine may have saved thousands of lives and billions of dollars.

Amid the aftermath of the Great East Japan Earthquake and the meltdown at Fukushima Daiichi, it came to light that not only were Okamura's prescient warnings ignored, but a startling number of other warnings from other potential Cassandras, dating back decades, were also tossed aside.

The one who was closest to Okamura in the accuracy of his prophecy is Katsuhiko Ishibashi. Ishibashi is a seismologist who in 1997 coined the term "nuclear earthquake disaster" in an eerily prescient paper. By then, he already had a history of watching his disaster predictions come true. His 1994 book, A Seismologist Warns, criticized Japan's lax building codes, which he argued put Japan's cities at risk. After an earthquake killed thousands of people in Kobe less than half a year after the book's publication, A Seismologist Warns became a best seller.

From a young age, Ishibashi feared earthquakes. As a child he slept with a flashlight at his side in case an earthquake struck and he needed to escape quickly. In 1964, Ishibashi was in college when a magnitude 7.5 quake shook his apartment in Tokyo. Radio coverage of the event mentioned that Japan did not have enough earthquake experts, and Ishibashi took note. He decided he would become one.

Ishibashi first became concerned about how nuclear plants would fare in severe earthquake scenarios in 1995 when a magnitude 6.9 quake destroyed parts of an expressway and killed more than five thousand people. In the *International Herald-Tribune*, he wrote an article that, unfortunately, would later seem prescient. He prophesied that a significant earthquake could take out reactors' external power, that a tsunami could then overtop the seawalls, flood the EDGs, disable cooling of the reactors and lead to meltdowns.[9] If this sounds very familiar, that's because it is exactly what happened at Fukushima Daiichi in 2011.

Experts dismissed Ishibashi's article. Haruki Madarame, who was then teaching at the University of Tokyo School of Engineering and would later be chairman of Japan's Nuclear Safety Commission (the position he held at the time of the Fukushima disaster), wrote in a letter to the Shizuoka legislature that Ishibashi was "a nobody" in the world of nuclear engineering. On March 12, 2011, Madarame became an object of ridicule and revulsion in Japan for reportedly telling Prime Minister Naoto Kan, just a few hours before a hydrogen explosion blew up Fukushima Daiichi Reactor 1, that there would not be an explosion there.

Other warnings were also ignored, including one that came out of a 2000 TEPCO in-house study. The study acknowledged the possibility of tsunami waves of up to fifty feet, a height that would overtop Fukushima Daiichi's nineteen-foot seawalls. It recommended that measures should immediately be taken to provide better protection from potential seawater flooding, measures that were obviously never adopted.

TEPCO headquarters consistently insisted that the risk was unrealistic. They brushed off the fifty-foot figure warned of in the 2000 report, saying that at the time "the technological validity could not be verified." A TEPCO report after the catastrophe struck acknowledged that "when deliberations were held internally about the results of a tentative calculation concerning a tsunami strike exceeding assumptions at [fifty feet], said risk was not announced" because of difficulties communicating "risk information." There was a fear that "announcing information about uncertain risks would create anxiety" in the surrounding communities and "result in a decline in capacity utilization rates."[10]

Put simply: if we are honest about the risks, the thinking went, the locals may protest and refuse to buy our energy in an attempt to force us out of their towns. This rationale, as we will see, was a major factor in why warnings regarding the risk of tsunamis were ignored by energy and government executives in Japan more generally.

HOW COULD SO MANY CASSANDRAS HAVE BEEN IGNORED FOR SO LONG?

Okamura's warning was directly about Fukushima, and the fact that it was unheeded tears at him. "I think [TEPCO] could have reduced the death toll and saved some lives," he lamented. "But they failed to do so in time. I am extremely sorry, and it is a shame."[11]

The level of human failing was so vast, the degree of arrogance so profound, and the lack of preparation so remarkable, said the 2012 Fukushima Nuclear Accident Independent Investigation Commission, that it considered the disaster to be "man-made." Inevitably, the question arose, "Why?" But to understand the answers, we first must understand a bit more about the broader history and context of Japan's nuclear industry.

Japan's geology is both what initially drew the country to the promise of nuclear power and what makes the country such a dangerous place to use it. With few fossil fuel reserves in the ground, Japan has historically imported most of its energy. Relying on nuclear power gave Japan a way to limit the amount of money it had to spend externally.

Before the Fukushima accident, nuclear power provided 30 percent of Japan's energy. The government intended to up that number by 10 percent. But Japan's position on the Pacific Ring of Fire, a highly active seismic area, makes it very prone to earthquakes. In fact, 90 percent of the world's largest earthquakes and the vast majority of volcanic eruptions occur along the Ring of Fire. Japan itself is home to about 10 percent of the active volcanoes in the world and experiences up to 1,500 earthquakes per year.

In light of this geological instability, nuclear power in Japan presents some higher-than-average risks, but because of its limited supply of energy resources, the country was willing to take them. Japan's reliance on nuclear power gave the Japanese government and the nuclear industry reason not only to work against the country's strong emotional reactions to nuclear power borne from the bombing of Hiroshima and Nagasaki, but also to play down or ignore warnings regarding safety risks at nuclear power plants.

This reason to rationalize helps to explain how the prescient warnings of Dr. Okamura and others were ignored, and why Fukushima tragically became such a costly Cassandra Event. We postulate that there were four main reasons: the willingness by TEPCO and the Japanese government to accept an unusually high level of risk; a myth of nuclear safety pushed by the Japanese government; regulatory capture, the collusion of power companies and nuclear regulators; and, as we have seen in each preceding case, Initial Occurrence Syndrome.

First, TEPCO's risk tolerance in building and operating its nuclear facilities was apparently very high. Extensive efforts to conceal safety risks, going back decades, were exposed following the disaster. It became clear that although the events at the power plant had been set into motion by natural causes, human negligence and indifference played a large role in creating the conditions for the nuclear tragedy.

The Fukushima Nuclear Accident Independent Investigation Commission faulted both TEPCO and government agencies for failing to develop needed safety requirements. It found "ignorance and arrogance unforgivable for anyone or any organization that deals with nuclear power." For decades, TEPCO had tamped down possible safety concerns by downplaying or concealing information to save money, avoid inconvenience, and keep such matters out of the public eye. Before the Great East Japan Earthquake, fourteen lawsuits had been filed accusing TEPCO of ignoring or hiding risks to avoid spending money. None of these trials, however, led to any change. Nuclear power companies in Japan had typically received little or no punishment for their safety violations; regulators had long turned a blind eye.

TEPCO did practically nothing in response to the warnings from any of the Cassandras. As Okamura put it in a biting understatement, the Fukushima disaster "was partly a natural phenomenon, but it is true that there had been various warnings. The opinions of experts like me have not necessarily been taken as important."[12]

TEPCO later admitted that it had failed to develop safety requirements out of fear of possible lawsuits or protests. This motivation is linked to the second broad reason why so many warnings were ignored: TEPCO's effort (as well as the Japanese government's) to maintain a positive public opinion of the nuclear industry. Such a positive attitude was especially hard-earned, given the fact that the 1945 attacks on Hiroshima and Nagasaki, which killed hundreds of thousands of people, had a hugely traumatic effect on the country.

When they began the nuclear power industry, Japanese leaders, under considerable influence from the United States during the "Atoms for Peace" initiative to develop peaceful uses of nuclear power, hoped to build the first plant right in Hiroshima. The Japanese leadership reasoned that it would be much harder for the rest of Japan to oppose nuclear power if Hiroshima itself embraced it. That idea was not implemented, but in the end, the country's desire to advance technologically and limit its reliance on outside energy sources trumped the emotional opposition to nuclear power.

Earlier Japanese leaders had created what Prime Minister Yoshihiko Noda called a "safety myth," which led people in Japan to believe that nuclear technology could not fail. Perpetuating the myth motivated the government and the power companies to keep private whatever dangerous scenarios they might have been considering.

The safety myth helped to enable significant collusion among companies like TEPCO and nuclear regulators, the third major factor in explaining why they ignored the warnings: regulatory capture. The report by the Fukushima Nuclear Accident Independent Investigation Commission, the product of a six-month investigation, details the relationship of complicity and collusion between TEPCO and its regulator, NISA, to avoid any changes to TEPCO's plants in order to comply with new safety guidelines. The report calls the product of this collusion "regulatory capture," which mostly meant that the foxes were watching the henhouse. Lack of clarity concerning the roles of utility companies, nuclear regulators, and the rest of the government, compounded by poor communication among them, only deepened the problem.

As physicist Amory Lovins succinctly explains, Japan's "rigid bureaucratic structures, reluctance to send bad news upwards, need
to save face, weak development of policy alternatives, eagerness to
preserve nuclear power's public acceptance, and politically fragile
government, along with TEPCO's very hierarchical management culture"[13] all played a role in the disaster. It was not simply that Okamura
raised his warning at the wrong time or that he hadn't strenuously
demanded a meeting with NISA to consider the risk from tsunamis.
(TEPCO claims that further investigation was, in fact, underway at
the time the Great East Japan Earthquake hit.) While the earthquake
itself was the result of 1,142 years' worth of stress built up between two
lithospheric plates, the actual humanitarian and environmental disaster was the result of years of bureaucratic concealment and collusion.

A fourth major factor in TEPCO and the regulators' refusal to
respond to any of the many warnings is yet again the availability bias,
and more specifically, Initial Occurrence Syndrome. The 869 Jōgan
Jishin earthquake and tsunami seemed too distant to even be considered. Had more people in Japan been aware of the destructive force
of that millennium-old event, or had the Active Fault and Earthquake
Research Center's new studies received more public attention, perhaps people would have been able to accommodate the idea that
such a thing had indeed happened and might happen again.

With availability bias shielding regulators from considering the
historical record and the pervasive myth of nuclear safety in Japan
preventing a deeper examination of the risks, compounded by col-
lusive regulatory capture, a perfect storm of conditions existed under
which perhaps no Cassandra stood a chance. The expertise of people
like Okamura and Ishibashi had little power against the cultural belief in nuclear power's safety and the bureaucratic arrangements that
concealed so much risk. Indeed, it is difficult now to envision ways
that circumstances could have been altered to create a scenario in

which Okamura's warnings might have been heeded. Over the years, regulatory capture and TEPCO's prioritization of public acceptance had defeated progress on safety guidelines.

After the Fukushima disaster, Japan's nuclear industry quickly lost support. Decades of concealment by the industry and its regulators of the dangers of nuclear power in this seismically active zone had backfired dramatically. A few months later, 75 percent of people polled in Japan reported that they were against nuclear energy. More than 80 percent did not trust the information coming from the government or TEPCO regarding the disaster. Trillions of yen had been spent building nuclear power plants. Now they lay dormant, unlikely to be reactivated anytime soon. Japan was again reliant on imported fossil fuels to generate electricity.

The TEPCO officials and the rest of the bureaucratic leaders involved will forever have to live with the knowledge that they ignored Yukinobu Okamura, the polite but insistent man who tried to alter the tide of history for the better. Okamura will forever be haunted by the thought that perhaps if he had done more, if he had spoken out more loudly, things might have turned out differently. "Now I regret that I didn't stress this more strongly, to push them to research this," he reflected.[14]

The Fukushima meltdown, a natural disaster that turned into a $100 billion nuclear tragedy, could have been prevented if the warnings of a quiet scientist named Yukinobu Okamura had been heeded.

CHAPTER 6

The Accountant:
Madoff's Ponzi Scheme

Lesson number one: Don't underestimate the other guy's greed.

—FRANK LOPEZ, *SCARFACE* (1983)

A snub-nosed .38 revolver, the kind of gun that Jack Ruby used to kill Lee Harvey Oswald in the basement of the Dallas police headquarters, is an old-school handgun, a relic, the stuff of 1940s mobsters in fedoras, but it's reliable and shoots straight. It will do just fine in a close ambush where you have to move fast and it's all over in a shot or two. After all, that's the kind of situation Harry Markopolos figured he'd be facing if somebody was going to be coming after him.[1] Markopolos checked the cylinder, snapped it home, and tucked the revolver back into his holster. He was ready for the day.

Harry Markopolos was a certified financial advisor, working as a portfolio manager for a good-size investment company in Boston. Men in his line of work do not normally need to fear hit men, but there was nothing normal about where Markopolos found himself. He was the leader of a small group of finance professionals who were secretly investigating Bernie Madoff. They knew Madoff was running a secretive, exclusive, money-management service, with clients ranging from Palm Beach princesses to European royalty to Russian mobsters. But Harry also knew Madoff was nothing but a hollow sham.

In 2008, Madoff was arrested and confessed to the FBI. After the bottom fell out, $65 billion vanished overnight. Thousands of investors were wiped out—lost their retirement savings, forced to sell homes, flung overnight from security to desperation. Three people, if not more, died violently.

But this was 2002, and Harry Markopolos was desperately trying to convince the Securities and Exchange Commission (SEC) to bust Madoff. For two years, Markopolos had been trying to get the watchdog agency to do something. Then he would keep on trying for six more years, bombarding the SEC with letters, meetings, mathematical proofs. It didn't do any good.

Meanwhile, Madoff's elaborate fraud was still getting larger. His investors had already entrusted more than $12 billion to him. He was earning them great returns, at least on paper, reporting profits that ran into the hundreds of millions for some, even billions. And Markopolos was pretty sure some of them would kill to keep it that way.

THE CON MAN AND THE CASSANDRA

For Harry Markopolos, the story began in 1999. By that time, Bernie Madoff and his clients were already in the money. Big money. Madoff had a $7 million, two-story penthouse in Manhattan, three beach houses (Palm Beach, the French Riviera, and Montauk), and two sleek yachts. His wife liked to shop in Monte Carlo. He was named Chairman of the NASDAQ stock exchange. Not bad for a kid from Queens whose parents' small brokerage had been shut down by the SEC.

Bernie had married his high-school sweetheart, Ruth, and started his own brokerage in 1960 with $5,000 he had earned as a lifeguard. Ruth did the books. By 1999, he was a titan of Wall Street, a big broker-dealer, a middleman who buys and sells stocks, pairing buyers and sellers and making a few cents per share on the spread. By the early 1990s, his company was handling about 10 percent of the daily trades executed on the New York Stock Exchange.

Bernard L. Madoff Investment Securities also invested money for a select group of private clients. It was an insider thing, strictly word-of-mouth, a tip shared among friends and relatives. At the Palm Beach Country Club (entry fee: $350,000), maybe a third of the members had money invested with Madoff. His main feeder in Palm Beach, the guy you went see if you were hoping Bernie would open an account for you, was Robert Jaffe, son-in-law of one of Bernie's oldest friends and clients. People opened accounts for their grandkids. Synagogues, charities, and foundations learned about Bernie from their donors, and gratefully invested. His client list grew to include high-profile figures such as Steven Spielberg, Senator Frank Lautenberg, *New York Daily News* owner Mort Zuckerman, and L'Oreal heiress Liliane Bettencourt, the richest woman in France.

Harry Markopolos was at his desk on the twenty-third floor of an office tower in Boston's financial district, where he worked as a portfolio manager for Rampart Investment Management. Harry is not a particularly big guy, but he is intense, sure of himself, and simmering with internal energy. His brown hair is parted on the side, a little unruly. When he concentrates on something, his focus is like a buzz saw. Right at that moment, he was focusing on a page of numbers.

Markopolos likes numbers; he *gets* numbers. He can look at a page of tax returns and see the story they are telling. He also has a stubborn sense of right and wrong and does not like to back down. He went to a Catholic school and credits the Jesuits with teaching him to ask probing questions. The Army Reserve—he was commissioned as an officer, a rifle platoon leader, at age twenty-one—taught him to take initiative and trust his gut. His father owned a bar for a while, and then a growing string of Arthur Treacher's Fish & Chips restaurants, where Harry uncovered his first fraud involving an employee who was filching cartons of frozen fish fingers. Later Harry went into finance and ended up specializing in complex derivatives. "Numbers don't lie," he repeatedly told us.

Harry shared his office at Rampart with Neil Chelo, a young analyst he was training in quantitative analysis. Their desks were pushed together, face-to-face. He was talking with Frank Casey, who had a knack for explaining Rampart's complex financial products to customers. Harry had heard from a friend at a Madison Avenue hedge fund called Access International Advisors that they put a lot of money with a manager who was getting their clients 12 percent a year or better, with returns of 1 to 2 percent net per month, consistently, month after month. With that level of reliability and safety, this manager was their hands-down favorite. Access was essentially just pooling its clients' money and handing it over to him to manage. Then Access just sat back and collected fees from their clients. Access was, in essence, a feeder fund, and it was making good money that way.

To find out more about this impressive hedge fund, Rampart sent Frank Casey to meet with the head of Access, an urbane French financier named René-Thierry de la Villehuchet. Frank Casey had come back with surprising news, de la Villehuchet shared his secret: the mysterious wizard was Bernie Madoff. Madoff's company was known to everyone as a broker-dealer, but none of them had ever heard that he was taking in private investment and managing money like a hedge fund. Harry listened to the strategy that Madoff was supposedly using and looked at the figures. "It doesn't make any damn sense," he told Casey. "This has to be a Ponzi scheme."[2]

"It was obvious," Harry told us when we met with him in Boston. "It literally took five minutes" to see that Madoff was faking his returns.[3] The strategy that Madoff was supposedly using couldn't produce those kinds of results.

One of the few things Frank Casey had learned about Madoff's asset management operation from Thierry de la Villehuchet at Access International was that Madoff claimed to use an investment strategy called split-strike conversion. It was sophisticated enough that most investors would not be equipped to probe it too deeply, but it was not rocket science. Madoff would buy a selection of blue-chip stocks, then hedge his positions with put and call options, essentially financial insurance plans, to protect against losses. It's a safe strategy, but it does not have potential for huge returns. Harry had plenty of experience with similar products, and he knew that they were simply not capable of generating the consistently high returns that Madoff was reporting.

Markopolos started gathering as much information as he could on Madoff's performance record. He obtained monthly data for more than seven years and was astonished to see that Madoff reported only three down months in that entire period.

"I like to give a baseball analogy," Harry told us over lunch at the Langham Hotel in Boston's financial district. "If Bernie Madoff was a major league baseball player, he'd be batting .964 because 96.4 percent of his months were positive. He would rarely strike out, and never had back-to-back losing months. And that should have been your clue." Harry talks fast, with no pauses and no hesitation, like somebody who is certain about what he is saying and doesn't want to waste a lot of time. "It wasn't necessarily the returns that were the giveaway; it was the lack of risk and his *consistency* of returns." Even with an elaborate hedge of options trading, Madoff's strategy was based on buying and selling stock, and some months those stocks went down.

Madoff's portfolio could not *always* be up. What Madoff was offering his clients was not "the holy grail of investment products"; it was just simply impossible.[4]

Then there was the matter of Madoff's secrecy. If you were running one of the most successful hedge funds in the business, why wouldn't you advertise and start raking in profits? Instead, Madoff practically swore his clients and feeder funds to secrecy.

This was not a tiny operation that might understandably go unnoticed throughout the financial world; it was huge. In fact, Harry soon came to suspect that Madoff's asset management business was quite possibly the biggest hedge fund in the world. Tallying the funds they knew relied on Madoff as their money manager, Harry and his team came to estimate that Madoff had $3 billion to $6 billion under management. That was in 1999, when the biggest hedge funds in the world were managing about $2 billion dollars.

The scale raised a big, glaring red flag. A trading operation that large would leave footprints in the market, humongous footprints all over the place. If Madoff was managing as much as $6 billion, the number of puts and calls he would need to hedge his portfolio would be staggering. Yet there was no sign of such a giant player stalking the options markets, throwing around at least three times as much money as the next-largest hedge funds in the world. In fact, the number of options he would need to be buying and selling, Harry quickly figured, would far exceed the *total* number of options available in the world.

Whatever Madoff was doing, it could not possibly be a split-strike conversion strategy. But there was another explanation. It was simple and it fit the facts. Bernie Madoff was using a strategy like this: discreetly advertise your great results to attract a stream of new investors, and use that new money, flowing in by the millions, to pay out returns to your existing customers. For this strategy, you do not need to be a genius, working the market night and day; you don't need to make any trades at all. You just need to keep taking in new money so you have enough to pay off investors who want to cash out. All the rest, you keep.

There is a name for this strategy, and it is called a Ponzi scheme, named for the Italian-American Charles Ponzi, who ran a short-lived version of such a racket in 1920 in Boston, conning a flood of new investors. A Ponzi scheme involves no trades and no investments. It is a pure sham. And that is what Harry figured Bernie Madoff was doing.

THE CASSANDRA MOMENT

At Rampart Investment Management, Markopolos, Casey, and Chelo were fascinated by the scam. Unfortunately, however, their bosses were not interested in debunking Madoff's success; they wanted to imitate it. They told Markopolos to come up with a strategy that would produce the same fantastic results—high steady yields with no downside ever. Harry knew it was impossible, but his bosses were not listening. After about six months, he was sick of it. "I went to the SEC primarily for my own self-interest," he says in his 2010 book, *No One Was Listening*. "I wanted to rid myself of the pressure of having to develop a product that couldn't be created."[5]

In the spring of 2000, Harry worked late for a few weeks preparing an eight-page report to alert the SEC that Madoff was running a fraudulent investment scam of enormous proportions. He knew he was right, but like many Cassandras, he was conscientious about subjecting his own reasoning to additional scrutiny. He asked Neil Chelo to check his math, consulted with colleagues, and took his work to his mentor, a quantitative Jedi Master named Dan DiBartolomeo, whom Harry described as "a super-brilliant mathematician who . . . is a lot smarter than Neil or I will ever be."

Harry was in good spirits as he made his way to a small conference room in the SEC's Boston office in May 2000 to meet with
Grant Ward, the New England regional director of enforcement. "I
had given them the case on a silver platter and gift-wrapped it too," he
recalls now. But as he gave his presentation, he could see Ward's eyes
glaze over. "It very quickly became clear that he didn't understand a
single word I said after hello."[6]

Grant Ward was a lawyer. Even though Markopolos had been
careful to leave out all the complicated math, Ward just did not know
enough about hedge fund strategies and options trading to understand the presentation. Worse, he did not seem to have any interest
in learning. The Boston District Office decided not to pursue the
complaint or even forward it to the SEC in New York, where Madoff
was based. Harry was not just disappointed, he was stunned.

Markopolos and his little team at Rampart continued their unof-
ficial investigation, fitting in time around their actual jobs, without
their bosses realizing that they were working on something that could
shake the financial world.

In early 2001, they recruited a fourth member to their team: Michael Ocrant, a seasoned financial reporter who was the editor in
chief of a trade magazine called *Managed Accounts Reports* (MAR)
and its offshoot *MARHedge*. Casey and Ocrant gathered information,
learning of more and more investors and fund managers who were
placing money with Madoff. Markopolos and Chelo compiled the
data and crunched the numbers. As they pieced together information, the hidden whale was looking bigger and bigger, a real monster
in the depths.

It was too big to let it go, and because nobody else seemed to be
doing anything about it, Harry figured it was up to him to try again.
So in March 2001, Harry formally submitted a second warning to the
Boston office of the SEC. This time his report was three pages lon-

ger, with more evidence, another year's record of impossibly steady returns, and a better sense of the huge amount of money Madoff had collected from investors: more than $12 billion Markopolos now estimated.

Once again, no response from the SEC. Years later Markopolos found out that the Boston office did refer the complaint to New York, where the assistant regional director of enforcement spent a single day looking into Harry's carefully researched warning. She found that Madoff was not registered as an investment advisor—and therefore, by definition, could not be a corrupt investment advisor. "After reviewing the complaint," she concluded, "I don't think we should pursue this matter further."[7]

Bernard L. Madoff Investment Securities was located in the Lipstick Building, a slim oval of glass and glossy maroon stone telescoping up from Third Avenue in midtown Manhattan. The brokerage, the well-known public side of Madoff's business, occupied the whole nineteenth floor, with workstations for at least fifty traders. The firm was a lot bigger than when Madoff started it in 1960, but it was still a family business. Madoff's two sons, Mark and Andrew, oversaw the brokers. There was a big glassed-in office for Bernie, and a smaller one for his younger brother, Peter, who was senior manager of trading. Down a spiral staircase, on the eighteenth floor, was Ruth's office and the compliance department, where Peter's daughter Shana worked as in-house legal counsel.

Another level down, on seventeen, is where the magic happened. The traders were not allowed down there. It was modest, even dingy. Here was the home of Madoff's fraudulent investment service. An old IBM computer churned out monthly statements that were mailed to investors. Frank DiPascali, a college dropout who had worked for Bernie since 1975, presided over a staff of ten people with only modest clerical training.

In Palm Beach, there was Robert Jaffe, working the golf courses and beachfront tables at the Breakers. His operations were dwarfed by the Fairfield Greenwich Group, an investment firm founded by Walter M. Noel. Operating mostly from Greenwich, Connecticut, Noel belonged to country clubs and rubbed shoulders with the wealthy social set. Through family connections, Fairfield Greenwich mined overseas markets and the WASP-y lawn parties of Westchester and Greenwich, eventually becoming Bernie's biggest U.S. feeder fund. By 2001, it had turned about $3.3 billion over to Madoff.[8] Madoff also had an old family connection who tapped rich Jewish clients in Manhattan. J. Ezra Merkin was the chairman of lending giant GMAC and president of the Fifth Avenue Synagogue. A trusted pillar of the Upper East Side Jewish community, Merkin was known as "one of the wisest men on Wall Street," and he trusted Madoff with his clients' money.[9]

In April 2001, one member of Harry's team walked right into the heart of the beast. Mike Ocrant went to the Lipstick Building. Armed with everything that Harry knew, Ocrant interviewed Bernie Madoff. For every suspicious anomaly, Madoff had a breezy answer. Ocrant wrote an article for his magazine, *MARHedge*. Avoiding outright accusations, Ocrant ran through Harry's red flags: the impossibly consistent returns, the lack of trading footprint. Finally, Ocrant asked, what was in it for Madoff? Madoff charged no fees, contenting himself with broker's commission at the modest rate of four cents a share. Meanwhile, his feeder funds, which did nothing but funnel new money to Madoff, were charging 1 or 2 percent management fees and pocketing 20 percent of all profits. Under those conditions, if they were true, Madoff was the most unselfish guy on Wall Street.

Six days later, a similar story appeared in the financial magazine

Barron's, which has a much larger readership than *MARHedge*. Picking up on Madoff's irrational strategy of leaving all the profits to his feeder funds and investors, author Erin Arvedlund plugged in numbers and figured that Madoff was foregoing $240 million a year. Markopolos and his two teammates at Rampart were thrilled, thinking it would provoke an SEC investigation for sure. The SEC did nothing. Even at Rampart, nobody was listening. Harry's bosses cared only about finding a product that could compete with Madoff and capture some of the action.

Harry eventually came up with a new financial product for Rampart. He was not thrilled with it, but Rampart eagerly launched it, in partnership with Thierry de la Villehuchet's Access International. To help de la Villehuchet sell the product to his wealthy European clients, Harry went to Europe.[10] They met with fourteen hedge funds and private banks that they discovered proudly placed money with Bernie Madoff—and all fourteen believed that Madoff was accepting new money only from them. He was quietly taking in a lot of new money but hiding it from other investors, a telltale sign of a Ponzi scheme.

Harry's European tour also gave him a glimpse of a more menacing secret: a lot of these funds that were investing millions with Madoff were offshore funds. "The *best* people in an offshore feeder fund are going to be tax cheats, and it's going to go quickly downhill from there to organized crime," Harry explained to us, amid the civilized tinkle of cutlery and crystal at the Langham. "So I realized that Bernie was stealing from the Russians and the drug cartels." Which meant that the man who was gunning to take down Madoff's money machine suddenly had enemies a lot more dangerous than he had ever imagined. That's when Harry started carrying that snub-nosed Smith & Wesson .38.

"WE DON'T QUIT"

Harry and his wife welcomed twins into the family in 2003. Frank Casey had left Rampart in 2001; in October 2003, Neil Chelo left too. Mike Ocrant left journalism that same year. It seemed that everyone was moving on. The next year it was Harry's turn. Finance, he had decided, was a dirty business—rife with dishonesty, ethical shortcuts, and outright fraud. He had known that for a while, but tracking Madoff had driven it home. He wanted to be one of the good guys. In August 2004, he became a professional fraud investigator.

It had been four years since Harry first approached the SEC. Now he was out of the investment business and his anti-Madoff team was scattered. "But our investigation never stopped; it never even slowed down."[11] They kept tracking Madoff, asking questions, gathering data. "I came from an Army background, and the one thing I can say about the Army is that we don't quit."

In June 2005, their vigilance paid off: they picked up a first hint that Madoff was in trouble. Bernie was trying to get loans from banks and was getting refused by some. They suddenly considered Madoff a bad investment, and that meant that they knew something was up. Evidently new investors were not rushing in fast enough anymore. The economy was slowing down, and Madoff was starting to feel it.

Fear of Bernie's imminent collapse is what pushed Markopolos to make his third submission to the SEC, in fall 2005. He drew up another report. Under new management, the Boston SEC office put Harry in touch with the agency's New York regional office, which had jurisdiction over Madoff's Manhattan-based operation. Harry submitted his twenty-one-page report to Meaghan Cheung, one of the branch chiefs at the New York office, in November. The six red-flag warnings from his first submission, back in 2000, had now multiplied to thirty. By now, he estimated, Madoff's huge scheme had taken in $20 to

$50 billion. Madoff's 12 percent annual average returns now spanned fourteen and a half years, with only seven down months. The title of Harry's report was "The World's Largest Hedge Fund Is a Fraud." He was hoping the SEC would not miss the message this time.

The SEC did nothing for a full year. Finally, in January 2006, the New York officials opened an investigation, but they rejected Harry's offer to explain his findings or help them with the investigation. He drew up a list of banking and investment professionals for Cheung and her team to talk to. She never even looked at the list. Each time he called Cheung to offer help or check on the status of his complaint, Harry felt that she was dismissive and disdainful. Eventually he stopped calling. He concluded that the SEC staff was simply not willing to believe that Bernie Madoff could be a scammer. "If he was a fraud, it brought into question everything these people believed in. Bernie Madoff was the ultimate insider; I was the bothersome outsider. I was some quant [mathematics geek] from Boston nobody had ever heard of."[12]

Cheung's team went to the Lipstick Building and talked to Madoff. They found his behavior suspicious. They detected that he was lying to them on certain points, but they didn't challenge him. They determined that he had violated some SEC rules but decided that "those violations were not so serious as to warrant an enforcement action." In the end, they reported, "The staff found no evidence of fraud."[13] Instead, they concluded that Madoff was in violation by acting as an investment advisor without being registered as one. So he registered. And that was it. Case closed.

In June 2007, Harry offered his fourth submission to the SEC ("Hope dies hard," he wryly noted later.[14]) It was an update, presenting Cheung the results of his latest research and advising her that Madoff seemed increasingly desperate in his scramble to take in new money. "When Madoff finally does blow up," he warned, frightened investors would stampede toward the exit, whether or not they had money with Madoff. "It's going to be spectacular and lead to massive selling by hedge funds."[15]

Harry's latest warning, like the previous ones, did no good. And by that point, time was running out. When the 2008 financial crisis hit, investors started to bail out. That fall, Madoff was swamped with $6 billion in redemption demands from investors. He didn't have the money.

Madoff was stressed out, at the end of his rope. Sometimes he would just sit in his office and stare off into space. The people around him began to notice. On December 10, Mark and Andy went to their father's office to ask him what was going on. He said he couldn't talk about it there, he "wasn't sure he would be able to hold it together." They left the office together and went over to Bernie's penthouse. They stood in the kitchen. Ruth was there too. "It's all just one big lie," he told them. The hedge fund operation was "basically a giant Ponzi scheme." He was getting ready to turn himself in.[16] After Mark and Andy left the apartment, they called a friend at a top New York law firm, who alerted the SEC and federal prosecutors.

The next morning, at around eight a.m., Bernie Madoff answered a knock at the door to his apartment. In his pajamas and bathrobe, he opened the door to two FBI agents. "We're here to find out if there's an innocent explanation," the lead agent asked. "There is no innocent explanation," Madoff answered. He then confessed.[17] The agents arrested him and drove to the prosecutor's office for the Southern District of New York, where federal prosecutors charged him with securities fraud.

THE $65 BILLION DISASTER

The news spread almost instantly. In Palm Beach, which probably had the highest concentration of Madoff investors anywhere in the world, cell phones rang and tanned faces went white.[18] In the New York office of Fairfield Greenwich, salesmen looked up in horror as the news of Madoff's arrest came over the Bloomberg terminals; Fairfield Greenwich and its investors had just lost $7 billion. Ezra Merkin, the wise rabbi of Wall Street, had earned $470 million in fees by sending $2.4 billion of his clients' money to Madoff. Yeshiva University lost over $100 million. Elie Wiesel's Foundation for Humanity lost $15 million, everything it had; Wiesel and his wife lost millions more personally.[19] When it was all said and done, the collapse of Madoff's Ponzi scheme brought crippling financial losses to more than 13,500 victims.

The shock waves widened far beyond the individual account holders. When philanthropists and foundations took a hit, so did the hospitals, museums, and charities they supported. Thanks to his network of feeder funds, Madoff had built the first truly worldwide Ponzi scheme, with victims in Latin America, China, Japan, Korea, Singapore, and Dubai. Europe was especially hard hit by L'affaire Madoff, as it came to be called in Geneva and Paris. The middleman between Madoff and many of his European clients was Access International, which lost about $1.4 billion, including $55 million belonging to Thierry de la Villehuchet. By all accounts an honorable man, de la Villehuchet was mortified that he had led so many friends and associates into disaster. On December 22, he went to his office, wrote notes to his wife, his brother, and his business partner, took sleeping pills, and slit his wrists with a box cutter.

In 2009, Bernard Madoff admitted to eleven counts of securities and investment fraud, theft, perjury, and money laundering. Insisting that he was solely responsible for the fraud, he refused to cooperate with prosecutors investigating the possible involvement of his family or employees. He was sentenced to 150 years in a federal prison.

Madoff's sons, Mark and Andy, were never charged with criminal offenses. Bernie was presumably relieved, but not for long. On the second anniversary of his arrest, his older son, Mark, hanged himself in his New York apartment. Four years later, in September 2014, Andy died of mantle cell lymphoma, a cancer that spreads throughout the bloodstream. He was forty-eight.

Today Bernie Madoff lives in Butner medium-security penitentiary, in North Carolina, in an eight-by-ten-foot cell. With both his sons gone and Ruth no longer visiting, the loss of his family is reportedly the hardest blow. Madoff seems to have little remorse for his victims. "It's not like I ever considered myself a bad person," he told a reporter in 2014. "I made a horrible mistake and I'm sorry."[20] Other prisoners respect him for the grand scale of his operation. They ask him for investment tips all the time.

"DEAF, BLIND, AND MUTE"

Why did so many trust Madoff with their money? Psychology can trump rational decision making, and it is not hard to see some of the factors involved: the assumption that an established Wall Street insider had to be safe; the trust (intentionally) inspired by the insiders-only, friend-of-a-friend method of joining and the lure of exclusivity; and a very human desire for people to believe that they had found a financial wizard who could guarantee returns.

But what explains the SEC's extraordinary willingness to give Madoff the benefit of several doubts and dismiss specific, repeated, and compelling warnings from a well-informed, persistent Cassandra? "The government agency charged with being the industry's watchdog was deaf, blind, and mute," Harry wrote after the dust settled.[21]

Its failings were partly due to systemic shortcomings. Despite what people think, the SEC is not like an FBI for chasing financial crimes, stocked with well-trained and highly motivated agents empowered to storm into companies and turn the place inside out. In reality, the SEC has very limited authority to investigate and is relatively small and underfunded, particularly when compared to the industry it regulates.

In an industry awash with gigantic pay packages, the SEC's staff make humble government salaries. Naturally, the smartest, top-of-the-class candidates typically gravitate toward jobs within the industry. Accordingly, the SEC suffers from the same revolving-door problem that afflicts many government regulatory agencies: people work there for a short time before crossing over into the very companies they used to regulate. During his years of chasing Madoff, Harry became convinced that the SEC "had been captured by the private industry it was created to regulate."[22] As we saw in a preceding chapter, the same sort of regulatory capture also blinkered regulators prior to the Fukushima disaster.

Still, Harry doesn't believe corruption was the reason his warnings went unheeded. Instead, he believes the agency was simply riddled with systemic failures and incompetence, beginning with a staff that couldn't believe that such an established, well-known money manager could also be running a vast Ponzi scheme. "Size alone gave Bernie that patina of respectability," Harry pointed out to us. "Everybody at the *Wall Street Journal* and the SEC, along with the victims, assumed he was so big and there were so many big banks marketing his product, that he had to be legit because certainly they would have checked."

We actually know a lot about exactly how the SEC fumbled so badly because, after Madoff confessed and the agency's failure was made public, the SEC's own inspector general conducted an investigation and in September 2009 issued a 477-page report. The cycle of repeated warnings and failures highlights a variety of lapses with lessons about the ways that warnings can be wasted.

One of the biggest problems was what we will call Complexity Mismatch. As Harry explains, the SEC was employing the wrong *kind* of people: its staff was dominated by lawyers, not finance professionals. They didn't have the training to understand complex financial strategies, options, and derivatives. Instead of grizzled old industry veterans who had seen it all and learned the tricks, the investigators were mainly young lawyers whose keenest interest in the financial industry was to eventually get a job there.

Moreover, the lawyers at the SEC were looking for legal proof—documentary evidence that they could use in court—but what Harry brought them was mathematical. "Math is truth, finance is bullshit," Harry likes to say. He knew Madoff's numbers were impossible, and it was proof enough for anyone versed in finance and accounting that something was amiss in Madoff's operations. But his audience at the SEC simply couldn't understand the calculations, considering them circumstantial evidence at best.

The ineptitude of the SEC staff to police the complex financial industry was complicated by their ready acquiescence to "satisficing" solutions. We will discuss the term in more detail in chapter 9, but briefly, "satisficing" happens when a decision maker addresses the issue but doesn't solve the actual problem. For example, the IG's report notes the reason that the SEC's New York office didn't look into Markopolos's second complaint in March 2001: because Madoff wasn't registered as an asset manager, the SEC decided that the claim that Madoff was a *fraudulent* asset manager was without merit. Satisficing solutions are often simply window dressing, covering up the real issue and allowing the decision maker to tend to other, less problematic issues.

After Harry Markopolos submitted his third warning in the fall of 2005, the IG report notes, his complaint was investigated by a team that was woefully unprepared to investigate a Ponzi scheme. The team didn't understand even the basics of options trading or how a hedge fund was supposed to work. The main investigator, Simona Suh, was a young lawyer who had been with the agency less than two years. Because of their lack of expertise, the team essentially relied on Madoff's firm to provide them with data detailing their operations. The IG's 2009 report states, "The Enforcement staff's investigative plan primarily involved comparing documents and information that Madoff had provided to the examination staff (which he fabricated) with documents that Madoff had sent his investors (which he also fabricated)."[23] That is like conducting a murder investigation by asking the suspect to let you know if he did it.

The investigators were in over their heads. Despite having "almost immediately caught Madoff in lies and misrepresentations," they did not bother to press him or dig deeper, the SEC acknowledged in its internal review. "When Madoff provided evasive or contradictory answers to important questions in testimony, they simply accepted as plausible his explanations."[24] When asked, Suh could not provide a good reason why they had been so willing to take Madoff at his word. Madoff just "didn't fit the profile of a Ponzi schemer," she told the IG in 2009.[25] Unable to put all of the pieces together, the SEC conducted a satisficing investigation so that it could say it had done something when it effectively had done nothing at all.

While Suh took Madoff's comments at face value, Markopolos found that his message was received quite differently. In fact, his warning was complicated by a factor common to many Cassandras we have studied: the Cassandra's off-putting personality. When the SEC rejected Harry's 2005 warning about Madoff, personality was a powerful factor. Meaghan Cheung, the branch chief in New York, "took an instant dislike to Markopolos, and declined to even pick up the 'several inch thick folder on Madoff' that Markopolos offered," as the SEC's 2009 inquest acknowledges. Cheung's colleagues recalled that she was offended that Harry thought the SEC investigators might need his help to understand the case; her rapport with him became "adversarial."[26]

Harry identifies some failings in his own communication as partly responsible. "I was communicating with lawyers in the wrong fashion," he told us in Boston, noting that he should have simplified his report even more. In our very fascinating time with him, it was obvious how one could see Harry as a highly confident, even strident idealist, and one not interested in making friends or allies. He has called the SEC "a bunch of idiots" and said, "They were comatose. They didn't respond to heat and light, much less evidence of wrongdoing."[27] Describing one meeting with an SEC lawyer, Harry noted that the lawyer looked like he wanted to leap over the table and strangle him. Even committed investigators might find it hard to ally with such a forceful personality, let alone believe something too complex for their training and wildly more sinister than any financial history or reasonable expectation. The SEC staff made a fundamental error of human judgment by assessing the messenger and not the underlying message. Their distaste for Harry's irreverent style made them ignore the validity and urgency of his warning.

A SILVER LINING

One week after Madoff's arrest, President-elect Barack Obama cited the Madoff fraud in his announcement of his appointment of new SEC chairman Mary Schapiro. Coming at the same time as the stock market collapse of 2008, the $65 billion Madoff fraud showed the high price of not having effective financial oversight. Harry was quickly asked to testify before Congress.

"My goal was to make this the worst day in the entire history of the S.E.C.," Harry wrote soon afterward, "because the only way it was ever going to improve was to hit rock bottom."[28] He absolutely skewered the SEC, excoriating it for incompetence and timidity and asserting that it needed a wholesale staff turnover. A little over a month after his testimony, he met with the new head of the agency in its headquarters in Washington. "I showed her the statistics on whistleblowers versus law enforcement," he explained to us. "It was basically, 'Look, whistleblowers are twenty-three times more effective at detecting fraud than law enforcement.' And she said, 'We need a whistleblower program.' And shortly thereafter we had a whistleblower program, thanks to her efforts and leadership."

So is the SEC better now? "Night and day better," Harry says without hesitation. The new chairman implemented a number of Harry's suggestions. The agency set up a one-stop tip hotline, streamlined the process for launching an inquiry, and took on a record number of cases. It went after some of the biggest players on Wall Street. Goldman Sachs, for example, was slammed with a record fine of more than $500 million for fraud. Most important, Harry believes, was the new whistleblower program. Prior to the Madoff disaster, "there was no mechanism to come in." Now there is, and it's working. "They have the biggest cases they've ever seen with the most detailed information that they've ever seen," Harry tells us. "They have great cases walk through their door on a weekly basis."

The Madoff case made Harry Markopolos a star. A week after Madoff's arrest, the saga of his eight years of futile warnings was written up in the *Wall Street Journal*. Other newspapers quickly followed. He was on *60 Minutes*. His business as a private fraud investigator took off. In 2012, he collected his first reward from the SEC's whistleblower program. He told us that he is investigating numerous large frauds and stands to make a small fortune in future SEC rewards.

His experience with the rebuffed Madoff warnings taught him a valuable lesson about communication. "My cases tend to be big, in the billions. They tend to be global, transnational fraud schemes of a highly complex nature. So I have to simplify them so that the government can understand them. And I also have to simplify them for multiple audiences. I know I'm going to have attorneys in the U.S. Attorney's office or the SEC that are not going to have financial backgrounds, so I have to do things in picture book form and give them analogies," he explains. "But I also have to appeal to the forensic accountants who work largely with the opposite part of the brain, and give *them* something to sink their teeth into. So I have to be able to present the case across multiple dimensions to each audience that's going to be working on that case. That was something that I did not know how to do in the Madoff case, and so I have to blame myself for that."

Perhaps there were indeed characteristics of Markopolos's personality that contributed to his not being taken seriously, but the dismissive reaction he encountered is yet another example of a human bias that we must recognize can blind us to the truth. Society must rely on people like Harry Markopolos to see beyond the horizon and let us know when something is wrong, but as a Cassandra, Harry must rely on society to believe him. Have we all truly learned the requisite lessons to ensure that such a preventable disaster never happens again?

CHAPTER 7

The Inspector:
Mine Disaster

Mining is a dangerous profession. There's no way to make a mine completely safe: These are the words owners have always used to excuse needless deaths and the words miners use to prepare for them.

—TAWNI O'DELL

On Easter the mine was shut. The steep hills and secluded valley towns of southern West Virginia were covered in blooming redbud trees and brilliant forsythia, signaling the end of the long winter. Miners and their families dressed up a bit more than usual for church that day. Easter Sunday was special, and for twenty-nine of the miners, it would be their last.

Returning to the Upper Big Branch Mine early Monday morning, the miners noticed that the air was unusually still. Miners are always attentive to air flow, to ventilation. A few hours later, an explosion and fireball roared down those miles of tunnels, flaming forward at a thousand feet per second, scorching the mine and killing the miners. Joe Mackowiak had feared it would happen. He had tried to stop it.

In this chapter we examine another example of the federal government's attempts to institutionalize warning. Like the U.S. intelligence community's attempt to institutionalize warning by creating a national intelligence officer for warning, the government's efforts to set up a bureaucratic system to predict and prevent mine disasters had succeeded only in predicting, not in preventing. Joe Mackowiak was a key part of the federal government's attempt to create a Cassandra-like system, an agency to warn of impending disasters in the mines. Why did he see the disaster at Upper Big Branch coming when the company did not, and why was he ignored? How is it that, even when a calamity is predicted by government officials, the very people who presumably have access to the resources and authority needed to avert it, the tragedy occurs anyway?

What happened in West Virginia at the Upper Big Branch Mine began well before Easter weekend in 2010. The system and the disaster are part of a broader narrative of the U.S. mining industry that goes back more than a century.

A NATION'S PROSPERITY, AND PAIN

In 1742, an explorer named John Peter Salling made the first recorded mention of a substance that would come to inexorably shape the economy, politics, and culture of West Virginia in later centuries: coal. After the Civil War, demand for coal exploded as it powered the growing nation's rapid modernization and industrialization, and coal mining began in earnest in the state. In the mid and late nineteenth century, coal mining expanded from manual surface extraction to mechanized underground operations. Coal mining companies grew from small, local operations into major industrial players, expanding rapidly in size and complexity as well as power and influence.[1]

As efforts to extract coal reached deeper into the earth, the dangers grew. Miners, many of them teenage boys, worked underground with few safety precautions or working standards, lighting their way below the surface with open-flame oil cap lamps. Fires, cave-ins, and explosions from the ignition of highly flammable methane gas or coal dust were just a few of the fatal hazards they faced. The work was demanding, dirty, and dangerous.[2]

Mining has long been recognized as one of the most dangerous occupations, and it remains so even today. It seems logical, then, that this industry is one in which the government might attempt to create a system to identify signs of growing dangers in a mine's operation so as to warn of impending disaster and mitigate its severity. But coal companies, which had already grown politically influential by the end of the nineteenth century, opposed regulation that would have hindered coal production or cut into profits. So for the first decade of the twentieth century, there was essentially no oversight of the mining industry by the federal government, leaving to the private sector the responsibility of adopting and promulgating best practices in mining safety and health. Not surprisingly, this arrangement failed to protect the most vulnerable people in the industry, the coal miners.

The fatality rate in the coal mining industry climbed steadily through the turn of the century, and then held steady at around 34 deaths for every 10,000 miners per year (today the rate stands at about 1.4). The year 1907 opened with an ominous bang. January alone saw four major coal mining explosions in the United States, from Colorado to West Virginia. Two disasters claimed the lives of nearly a hundred West Virginians in that month alone. It was a horrific way to start the year, and people hoped that the worst was behind them.[3]

Then came "Bloody December," in which 692 miners perished in five separate mining disasters. The annual death toll already running well ahead of average, Bloody December brought the year's total to 3,242, the highest number of U.S. mining deaths ever recorded in one year. The greatest tragedy occurred at the Monongah Mine in West Virginia, where 362 men perished in a massive explosion, likely the result of a spark or open flame lamp igniting a cloud of methane or a buildup of coal dust. It remains the worst mining disaster in U.S. history.

The death toll of Bloody December, plus that of other mining disasters in the months that followed, resulted in a public outcry for a greater federal role in creating safety standards for the mining industry. Until then, coal companies had suppressed all attempts to create meaningful federal regulation for mine safety and had largely thwarted state efforts as well. Particularly in states whose economies were dependent upon coal, like West Virginia, safety laws were weak and seldom enforced. Still, after December 1907, coal companies knew that it was only a matter of time before some kind of federal action would take place.

Instead of fighting regulation head-on, they banded together and lobbied Congress to weaken the legislation under consideration. J. Davitt McAteer, former assistant secretary of labor for mine safety and health under President Bill Clinton and an expert on mine safety, recounted in his excellent investigative history of the Monongah disaster and in a later interview with us how the mine industry convinced Congress that the cause of the explosion was unknown and therefore it could not "prescribe medicine when they have not diagnosed the disease."[4] These efforts ultimately succeeded in convincing Congress not to pursue safety standards but to instead establish a research program on mining techniques, health, and safety.[5]

In 1910, Congress passed the Organic Act, establishing the U.S. Bureau of Mines (USBM, or to miners, simply the Bureau). The Bureau's mandate was "to increase health, safety, economy, and ef-

ficiency" in the country's heavy industries, including the mining industry. The USBM began by focusing its efforts on the problems that
had caused the vast majority of mining fatalities: the flammability
of methane gas and coal dust and the danger of explosives, electrical systems, and lighting systems nearby. Thus began a more than
century-long federal government effort to identify and mitigate mining dangers, an attempt to systematically create an institutional Cassandra that could warn of dangers and prevent future disasters.

In the years that followed, the Bureau publicized its research findings and collaborated closely with mine operators. This early legislation resulted in improvements in mining safety that reduced the
fatality rate by about a sixth, but the USBM had no authority to force
mine operators to adopt best practices and could not inspect mines for
dangerous conditions, a regulatory power that mine operators vehemently opposed. Hazardous conditions, like shoddy electrical wiring
and dangerous buildups of coal dust, remained a common sight. As
the Bureau lacked the authority to conduct inspections or enforce regulations, mine operators felt no need to enforce strict safety standards.
The USBM was incapable of serving as the industry's Cassandra.

By the end of the 1930s, the fatality rate in America's coal mines
had fallen to about 21 deaths per 10,000 coal workers, a reduction of
about a third since the start of the century. While the decrease can
be attributed to the Bureau's research and the voluntary adoption of
safety standards by some mining companies, coal mine disasters remained all too common.[6]

A series of major explosions in mines throughout the nation's
heartland in 1940 killed 276, the largest loss of life in one year in
mining disasters since 1928.[7] The events shook the nation and, like
the Bloody December of 1907, revitalized the grassroots call for better regulation of the industry and greater enforcement power for the

USBM. In response, Congress passed the Federal Coal Mine Inspection and Investigation Act of 1941. For the first time, federal law gave the Bureau its own authority to enter and inspect mines for hazardous conditions.

While the legislation threatened mine owners and operators with a misdemeanor charge for failing to open their mines to USBM inspectors, punishable by a fine and imprisonment, the Bureau still had no power to compel mine operators to comply with specific mine safety and health laws and regulations.[8] Steps to remedy any deficiencies found during inspections were considered mere recommendations rather than compulsory orders. Moreover, the companies remained potent political forces, particularly at the local level, and often used their influence to intimidate inspectors or weaken the enforcement of regulations.

Unsurprisingly, mine conditions and fatalities did not markedly improve in the years to follow. While the public was shocked in 1947 when a disastrous fire resulted in the deaths of 111 miners at Centralia No. 5 in Illinois, few directly involved in the mine's operations were surprised. Hazardous conditions were well known to federal and state officials, the mine operators, and of course the miners. In the years prior to this accident, inspectors documented numerous dangers, including excessive amounts of coal dust, dangerous roof conditions, elevated levels of carbon monoxide, and exposed electrical wiring. In a desperate plea for intervention, local miners even wrote a letter to the Illinois governor, begging him "to please save our lives."[9] Just a year later, their worst fears were realized. In later testimony before the legislature, one of the inspectors responsible for the mine said he feared the power and influence of the coal companies and believed he would have been fired for more aggressively enforcing regulations at the mine.

Just four years later and only fifty miles away, 119 were tragically killed when the Orient No. 2 mine exploded four days before Christmas, 1951. Like Centralia, inspectors uncovered dangerous deficiencies, including poor ventilation, methane gas, and dangerous dust conditions in the weeks and months before the explosion. But despite their reports' conclusions that the mine was "extremely hazardous," they had no authority to compel the mine operator to fix the problems.

After Centralia and Orient, the public again demanded action. President Harry Truman signed the Federal Coal Mine Safety Act on July 16, 1952. The new law was the first step toward putting real teeth behind federal safety and health regulation of the mining industry, giving the Bureau some actual power to prevent disaster. While the Bureau could not levy fines for specific deficiencies, it could order a mine closed if it suspected there was an imminent danger of disaster, and it could fine operators for failing to comply. The law also set specific requirements for mine ventilation and the suppression of coal dust, reducing two of the most potent causes of disastrous mine explosions.[10] The fatality rate in America's coal mines now began a long and sustained downward trend.

Still, death continued to plague the nation's mines. Subsequent years saw the same pattern repeat itself. Mining disasters spurred the development of new regulations, strengthening and expanding the scope of the institutional Cassandra system, but the deaths and injuries of the miners seemed to be the only thing that overcame corporate opposition and bureaucratic paralysis.

On November 20, 1968, a massive explosion and fire erupted at Consolidation Coal Company's No. 9 mine in Farmington, West Virginia, only five miles from the Monongah disaster of 1907 that first shocked the public's conscience. Futile rescue efforts were unable to reach the seventy-eight men trapped underground before nine subsequent explosions ripped through the mine, killing all inside. Like the surrounding hills of West Virginia that shook for miles, the public was once again shaken to action, and lawmakers were charged with bolstering Cassandra.[11]

The Coal Mine Safety and Health Act of 1969, the Coal Act for short, remains the toughest law ever passed in regulation of America's mines. Davitt McAteer, freshly out of law school at the time, was a part of the effort. When we asked how significant an accomplishment it was, he answered simply, "It revolutionized mine safety."

The Coal Act for the first time gave the federal government the ability to really enforce the law for the protection of miners. It established new health and safety standards for the nation's coal mines with substantial penalties for violations, and it required federal inspectors to inspect underground coal mines four times per year. But perhaps most significant, it enshrined in law the mantra that for too long coal companies had failed to live by: "The first priority and concern of all in the coal mining industry must be the health and safety of its most precious resource—the miner." In two years, the annual fatality rate in the coal industry fell below 10 deaths per 10,000 miners. It has never exceeded that threshold since.[12]

Coal miners have always known the risk of death, but other miners have also suffered devastating disasters. Shortly before noon on May 2, 1972, miners at the Sunshine Mine in Kellogg, Idaho, detected smoke, which subsequent investigation found was likely caused by the spontaneous combustion of waste materials and timber. A massive fire broke out and raged near the ventilation system's intake, rapidly filling the mine with smoke and carbon monoxide. Within hours, ninety-one miners were dead, many stuck deep within the mine when hoist operators were overcome.[13]

Sunshine was a silver mine and lacked the most dangerous hazards of coal mining, such as methane gas and coal dust. The disaster was a wake-up call for establishing a coherent set of safety standards covering the entire mining industry.

The subsequent 1977 Federal Mine Health and Safety Act, known as the Mine Act, realigned the U.S. government's responsibilities to enforce mine health and safety regulations under the newly formed Mine Safety and Health Administration (MSHA, pronounced "EM-sha") within the Department of Labor. The realignment eliminated a potential conflict of interest, leaving the Bureau of Mines solely in charge of conducting research on mine safety and efficiency. It also consolidated all mining regulations, for coal and all other mines, under the same law. The Mine Act remains the basis of the current federal inspection and enforcement regime.

By the dawn of the twenty-first century, the steady march toward improved safety standards and mine conditions resulted in the lowest miner fatality rate in U.S. history, fewer than 4 deaths per 10,000 miners. Sadly, each step of the progress in federal mining safety and health standards has come only on the heels of disasters, when public attention has galvanized Congress to pass more stringent laws and regulations.[14]

Most recently, the 2006 Mine Improvement and New Emergency Response (MINER) Act (again passed after tragedy, this time after twelve miners suffocated following a methane explosion at the Sago Mine in West Virginia) mandated improved emergency-response plans, procedures, and communications devices, as well as increased penalties for hazardous conditions. By then, the decades-long development of U.S. federal mining regulations and infrastructure had made huge strides in improving safety. In 2007, coal mining was statistically no longer the most dangerous industry in the nation, having been surpassed by agriculture, forestry, and fishing.[15] But despite the increasing empowerment of MSHA to inspect and enforce mine safety and health regulations, all had not been solved. Despite the system's proven efficacy, not everyone respected Cassandra's warnings. And the consequences would be disastrous.

"IF BASIC SAFETY MEASURES HAD BEEN IN PLACE . . ."

West Virginia is acutely aware of the dangers posed by mining operations. Since the turn of the twentieth century, it has had more fatalities due to coal mining than any other state: 2,617 deaths, more than a fifth of the U.S. total during that period. On April 5, 2010, it witnessed another tragedy, suffering the nation's worst mining disaster in forty years.

At 3:02 p.m., a powerful explosion tore through the Upper Big Branch (UBB) coal mine, operated by Performance Coal, a subsidiary of Massey Energy Company. The explosion propagated through the nearly two and a half miles of tunnels buried about a thousand feet below the surface of Raleigh County, West Virginia, with a force so strong that rocks, debris, and people near the entrances were ejected from the mine portals. Smoke and dust roared from the openings with a sound, described by witnesses, like a jet engine. Twenty-nine men who were working underground did not survive.[16]

To clearly understand what happened that day, it is necessary to take a closer look at the practice used at UBB known as longwall mining. Coal occurs naturally in striated bands within the Earth's crust, a layer of prehistoric organic matter compressed under heat and pressure for millions of years until it turns to rock. Natural gas, largely composed of methane, is formed by the same process and is trapped within and around the coal. Often coal deposits occur near enough to the surface to be dug directly out of the ground. In some cases, however, coal seams occur deep within the Earth and, as technology has progressed, deeper and deeper deposits have become accessible through underground mining.[17]

One of the most efficient ways to mine these buried layers of coal is known as longwall mining. First, miners utilize various machines and techniques to dig down to the target coal panel, the section of the coal seam they want to mine. Machines called continuous miners then cut parallel, horizontal tunnels along the sides of the panel, which are eventually connected by cutting another, perpendicular tunnel between the two, exposing the "working face" or "longwall face," from which the coal is mined.

To extract the coal, a massive electric cutting machine is used; it shears off a yard-thick layer of coal as it progresses along the working face, moving at about the speed at which you would take a casual stroll down the sidewalk. The shearer travels in one direction across the entire face, often for hundreds of yards, before reversing and shearing the longwall in the opposite direction. A series of hydraulic supports, called shields, support the mine's roof above the shearer. The shields automatically advance as the shearer cuts into the face, and the roof collapses behind them in a pile known as the gob. All the while, a series of conveyors transports the newly sheared coal to the surface and out of the mine.[18]

This process creates a significant amount of coal dust both in the air and on the surfaces of the mine. Research performed by the USBM in the early part of the twentieth century found that coal dust itself was dangerous enough to result in a massive coal mine explosion without the presence of explosive gases like methane, contrary to the belief of coal companies at the time. One of the most dangerous things about a coal explosion occurs when the pressure wave from the initial ignition disturbs nearby coal dust. This effect can create a propagating cloud of exploding dust that may travel for hundreds or thousands of yards. The Bureau found, however, that mixing incombustible rock dust with coal dust in the mine made it inert, no longer a danger to the miners. The long history of disasters resulting from coal dust explosions led to a federal requirement in 1952 that coal mines must maintain a mixture of at least 65 percent inert rock dust on mine surfaces.

Methane gas is often released from coal, the surrounding rock, and from the gob. While methane itself is not toxic, in high enough concentrations it can cause asphyxiation. Methane becomes even more dangerous in underground mines because of its violent flammability at concentrations above 5 percent in the air. Sparks can be created when a shearer's cutting bits come into contact with bits of ferrous rock intermixed with the softer coal along the working face. Worn down, improperly maintained cutting bits can leave "hot streaks," smears of extremely hot, nearly melted metal that result from the friction between the blunted metal bits and the rock. Either is capable of igniting a methane gas cloud, causing an explosion in poorly ventilated spaces. Consequently, shearers are equipped with water sprayers to suppress any potential sources of ignition while serving an additional function of circulating air and any methane that might be mixed in away from the cutting bits.[19]

Moreover, mines are required to maintain a certain amount of airflow throughout their spaces and past the working face. This is done specifically to ventilate and draw out any noxious or hazardous gases that are liberated from the mining operations and to prevent the formation of dangerous clouds of methane within the mine. The ventilation plan must be submitted to and approved by MSHA inspectors, who regularly inspect the mine to ensure compliance with this and other federal regulations.

The April 5 disaster at Upper Big Branch resulted from an unfortunate, coincidental, and completely preventable confluence of the hazards just described. The official accident report from MSHA declares, "The tragedy at UBB began with a methane ignition that transitioned into a small methane explosion that then set off a massive coal dust explosion. If basic safety measures had been in place that prevented any of these three events, there would have been no loss of life at UBB."[20]

Despite a century's worth of federal government efforts to institutionalize and facilitate a Cassandra system for the mining industry, conditions began to align that allowed the UBB disaster to happen. These conditions emerged slowly, over a period of months and even years as Massey Energy, the mine operator, willfully disregarded regulations and ignored safety violations. Unfortunately we were unable to interview Mr. Mackowiak for this book due to ongoing litigation between the U.S. government and Massey Energy. However, based on his testimony during the investigation and additional, extensive research, we feel it is still possible to adequately classify and discuss this as a Cassandra Event.

In 2008, Joseph Mackowiak was promoted to ventilation supervisor for MSHA's District 4, headquartered in Mount Hope, which covers southern West Virginia. He and his six engineers were responsible for ensuring that all of the 245 active underground coal mines in District 4 used plans that were compliant with federal regulations regarding how and how much air would be routed through the mine. In addition to approving these plans on paper, he and his inspectors also conducted complete inspections of each underground coal mine at least four times per year. Mines that are prone to giving off large quantities of methane gas, like the one at Upper Big Branch, received regular spot inspections because of the elevated danger of explosion. Given the large number of mines in District 4, Mackowiak and his engineers were kept very busy.[21]

Mackowiak had long worked for MSHA in mine safety, and even longer had been focused on the mining industry. Like many in West Virginia, Mackowiak's entire career has been devoted to coal. He began working for A. T. Massey Coal Company as a sophomore in college. There, he did engineering work for the company, which would later be renamed Massey Energy and whose subsidiary, Performance Coal, purchased and began operating the UBB coal mine in 1994. After graduating from college and working for Massey for several years, Mackowiak took a job with MSHA in September 2000, then later transferred to Mount Hope in 2004. He holds a bachelor's degree in mining engineering and a master's degree in safety, and has a long list of other specialized federal and state certificates and qualifications related to mining engineering, operations, and inspections.

From the moment Mackowiak became District 4 ventilation chief, UBB was a concern. The mine's ventilation plan was marginal at best: it was poorly engineered and unable to keep up with the changing internal nature of the mine. As Davitt McAteer explained to us, "Mines are constantly changing. There's a renewal every twenty-four hours, and that means the risks are constantly renewing themselves." As working sections are mined each day, as new sections are opened, and as old sections are closed, the internal dimensions and layout of the mine change. Each new day brings a new set of conditions, and hazards that didn't exist yesterday might have arisen overnight. In large, complex mines, these constant changes can have a tremendous impact on the dynamics of the air ventilating the mine, slowing or even reversing itself if a plan has not been well thought out. Such was the case at UBB.

Over the course of 2009, as his inspectors kept coming back with reports of new violations, Mackowiak grew increasingly concerned. These were not citations for minor, nitpicky details; they were citations for failing to get the basics right: not enough air moving through working sections of the mine, or air flowing in the wrong direction. He sent his ventilation specialists to the mine more often, sometimes even going himself, hoping he could keep close tabs on the conditions and whether or not Massey Energy was meeting its obligations to fix the problems he and his team had uncovered.

In September 2009, Mackowiak conducted a personal inspection of UBB, wanting to see for himself the ventilation conditions before the miners started developing a new longwall section. As he proceeded deeper into the mine, wind hit him in the face. According to the ventilation plan, it should have hit him in the back. Air flowing in the wrong direction could mean that dirty, dangerous air would be pulled across the hot, sparking longwall shearer or that noxious gases were simply recirculating within the mine. Either case was a serious hazard.

Mackowiak ordered that the mine be evacuated and shut down for four days until Massey could figure out how to restore the air flow. This was not an isolated occurrence. During testimony to the accident investigation board that was convened after the explosion, Mackowiak recounted the numerous ventilation plans submitted by the mine operator that he rejected for poor engineering or incomplete information, and the numerous times his inspectors returned with recurrent citations for failure to comply with even the most basic of ventilation safety procedures. Overall that year, MSHA inspected UBB more than any of the other 244 underground coal mines in the district, largely due to Mackowiak's concerns that it was a disaster waiting to happen.

Ventilation issues also were not the only problems to be found. UBB was a cesspool of other safety violations and hazardous conditions. In particular, coal dust remained an ever-present hazard.

Inadequate dust-suppression procedures meant that dangerously flammable amounts of coal dust existed almost constantly within the mine. The West Virginia Governor's Independent Investigation Panel, commissioned after the accident and led by Davitt McAteer, found that "in the 15 months preceding the disaster, UBB received citations from federal or state inspectors every month but one for rock dust issues. . . . Nearly half the 40 citations issued by MSHA were classified as 'significant and substantial.'"[22] For comparison's sake, over 10 percent of the violations issued at UBB were considered "unwarrantable failures to comply"; the national average is 2 percent.[23]

By March 2010, Mackowiak had become so alarmed by the reports of low air flow and air moving in the wrong direction that he took the unprecedented step of sending a team of five personnel to the mine so that they could measure airflow in different parts of the mine simultaneously. He had a sneaking suspicion that company supervisors, knowing an inspection was underway, ordered miners to divert air from other parts of the mine in an attempt to pass the inspections. Not surprisingly, the inspectors found problems: air was found traveling in the wrong direction through parts of the mine. The inspectors again ordered UBB shut down until the operator could correct the airflow issues.

A week later, yet another ventilation inspector found insufficient airflow in UBB. Alarmed and concerned by the report, Mackowiak called a former MSHA colleague, Bill Ross, who had taken a position at Massey Energy and who was working at a different mine, asking him to help remedy the problem. When Ross's superiors declined to transfer him, Mackowiak then took the very unorthodox step of contacting a senior Massey official directly. On March 16, 2010, he e-mailed Massey's vice president and chief operating officer, Chris Adkins, to request the transfer of Ross to UBB, noting the ongoing ventilation problems and asking for additional help. He never received a response.[24]

Just over two weeks later, operations at UBB shut down on April 4, 2010, for Easter Sunday. When the day shift returned early the next morning, they found the air down near the working face hot and still. Ventilation problems were nothing new to the miners, but this time, unbeknownst to them, something was different.

Methane had been leaking through the floor, accumulating behind the shields at the far end of the working face over the weekend, but with the shearer down for mechanical problems most of the shift, nothing seemed amiss. After shift change, the afternoon crew started up longwall operations as usual. Because crews were under pressure to constantly produce at the expense of safety and machine maintenance, many of the shearer's water sprays were inoperable, and its shearing bits were worn and blunted.

When the shearer approached the far end, a hot streak set alight a cloud of methane, starting a gas-fueled fire that burned for over a minute. As the crew cut the power to the shearer and began to evacuate, the fire ignited a cloud of methane that was mixed with just enough air to blow up. The explosion created a pressure wave that stirred up and ignited the adjacent coal dust, creating a propagating fireball that at times approached speeds of 1,000 feet per second and scorched miles of underground tunnels. Twenty-nine miners died as a result, ten from the explosion and fire and nineteen from carbon monoxide.[25]

INSTITUTIONAL REFUSAL, ET AL.

A century of U.S. federal mining regulation, written in the blood of miners, to create a Cassandra system that still failed to prevent disaster. How could this be? Why did such clear warnings go ignored? Does this mean that we cannot hope to employ such a system for warning? We believe the answer to the last question is no. The number of U.S. coal mine disasters dropped from 143 in the first decade of the twentieth century to 5 in the first decade of the twenty-first, while the amount of coal extracted annually more than doubled to over a billion tons. Federal regulation has played a critical role in making the industry safer and drastically reducing the frequency of disastrous accidents.

This Cassandra system's success is partly due to the research pioneered by the U.S. Bureau of Mines. As we have seen in the cases of other Cassandras, fundamental to their ability to make accurate predictions was their understanding of the underlying science, principles, and dynamics at work. Just as Dr. Ivor van Heerden's computer models of hurricane effects on the city of New Orleans and as Charlie Allen's knowledge of military history allowed both of them to accurately forecast the events to come, the USBM's development of mining safety knowledge provided the basis for a coherent and reliable set of warning indicators for the coal mining industry.

Over time, this same system also gave those charged with preventing disaster the ability to make their warnings heard. By the latter half of the twentieth century, federal officials had the ability to enforce mining health and safety regulations with fines, mine shutdowns, and even criminal prosecution. The 2007 MINER Act gave MSHA the authority to issue "flagrant violation" citations to willfully negligent mine operators and fines of up to $220,000 per instance as an additional tool to compel even the most recalcitrant of companies to deal with safety problems. Since the law was passed, the agency has used that authority more than a hundred times, totaling nearly $20 million in fines. But the authority was never invoked at UBB, despite the mine's poor safety reputation and ongoing concerns that the operator was not adequately fixing dangerous problems.

As we have seen, opposition to federal regulation by the coal companies has existed since the dawn of the industry. Mining companies maintained that they could regulate themselves and, as technical experts, were best suited to determine how to make their mines safer. Decades of horrific accidents and the unnecessary loss of thousands of miners' lives proved this assertion untrue and resulted in slow but significant steps toward improved safety standards, stricter regulation, and genuine enforcement power by the USBM and later MSHA. But some companies continued to ignore the arc of history, putting aside safety considerations in an attempt to maximize profits in a competitive industry where billions of dollars were at stake annually. Massey Energy was one of the worst offenders. "[Massey Energy] took the approach that they were going to push the envelope in every instance," Davitt McAteer told us bluntly. As an inspector, "if you go to a Massey mine, you know ahead of time they're going to come back and complain to you about a write-up. And they're going to complain to your boss."

MSHA had its hands full at UBB because the mine was not only poorly engineered, but the leadership of Performance Coal Company and Massey Energy also made it de facto company policy to put coal production and profits ahead of worker safety. Safety mechanisms designed to shut off mining systems if they detected elevated levels of methane were disabled. Security guards were trained to call ahead and give miners advance notice when MSHA officials arrived to conduct surprise inspections. Miners were threatened with retaliation for reporting safety concerns. Violations were contested and litigated rather than rectified. The UBB official accident investigation report found that Massey had created a "culture in which production trumped all other concerns."

Following the disaster, Massey Energy CEO Don Blankenship was charged with conspiracy to violate federal mine safety and health standards. For two months, jurors heard testimony from company officials, miners, and inspectors that Blankenship and other officials cut corners and turned a blind eye to safety in the name of production. In a memo to Blankenship that was introduced as evidence in the trial, a former MSHA official who had been hired by Massey Energy to help with ventilation and safety issues said he found that the company operated according to the idea that "we will run until we get caught, then we will fix it."[26]

Blankenship was found guilty of conspiring to violate federal mine safety standards, sentenced to a year in jail, and fined $250,000. Other Massey executives and supervisors were found guilty of similar charges: conspiracy to circumvent safety standards and procedures, evading MSHA inspections, and destroying documents and records. But their fines and prison sentences will never bring back the twenty-nine miners killed at UBB. Massey was not just negligent. It actively undermined the Cassandra system designed to warn and protect against disaster.[27]

Thus, in examining why Cassandra's warnings went unheeded, we first note that the disaster at UBB is an extreme example of institutional reluctance. Other Cassandras' warnings were ignored because the disaster they predicted had never before been seen: think of the U.S. government's reluctance to fortify New Orleans against a hurricane or the U.S. intelligence community's inability to believe that Saddam Hussein would invade Kuwait. In this case, the institution was not just reluctant to believe the warning; it refused to believe it. We may even call this extreme case institutional refusal. With institutional refusal, Cassandra faces one of the most challenging obstacles, because no amount of evidence can result in an effective response.

Without question, the responsibility for the accident and the deaths of the miners falls on the willful negligence of Massey Energy, its executives and leadership, who eschewed safety in the name of production and profits. But Cassandra was also given specific tools to use in the event she encountered institutional refusal, tools that were never fully used or even fully considered.

MSHA's internal after-action review of the agency's inspection and enforcement actions at UBB says that inspectors and officials broadly worked in good faith to enforce health and safety regulations. This was not an instance of regulatory capture, as in the Fukushima nuclear disaster. However, MSHA's report also says that inexperience, funding cuts, and personnel shortages contributed, at least in part, to the failure to pursue more aggressive policing at UBB. It found that some inspections were missed and that some of the most troubling inspection reports, like those that should have been considered for designation as "flagrant violations," were never reviewed by MSHA higher-ups.[28]

A second reason, then, that Cassandra's warnings went ignored is that the system's enforcement mechanisms were never fully utilized. Institutional reluctance existed not only on the part of those being warned, but also among the warners themselves. Additional numbers of experienced personnel may well have given closer scrutiny and levied greater penalties against Massey Energy, perhaps resulting in changes that would have prevented the April 5 explosion. But fundamentally, the system and the tools had already been established to facilitate a Cassandra's foresight and to ensure that the warning would be heard. Joe Mackowiak knew that the ventilation problems at UBB were serious enough to go outside the system and implore company officials directly, but he and his supervisors failed to fully appreciate that the worst could be imminent.

Yet while MSHA may have fallen short of using all the tools at its disposal against Massey Energy, the explosion at UBB that killed twenty-nine miners was a clear-cut case of warnings ignored, of Cassandra being dismissed. In general, the system fulfilled the intent of the 1969 Coal Act that "the first priority and concern of all in the coal, or other mining industry, must be the health and safety of its most precious resource—the miner." In numerous instances, it functioned as designed, flagging warning signs and prescribing corrective action. However, institutional refusal to comply was too high a hurdle to overcome. The disaster at Upper Big Branch unfolded over weeks and months. As Davitt McAteer explained, "April fifth wasn't the day all the failures at Upper Big Branch occurred. Ventilation. Water Sprays. Rock dust. Coal dust. Everything was already geared to collapse. It was only a matter of time."

CHAPTER 8

The Market Analyst: The 2008 Recession

But land is land, and it's safer than the stocks and bonds of Wall Street swindlers.

—EUGENE O'NEILL, *LONG DAY'S JOURNEY INTO NIGHT*

In August of 2007, Meredith Whitney overheard a comment at a cocktail meet-and-greet for financial analysts at Citigroup's New York City headquarters that inspired her to do something that shattered the world's financial markets.

Whitney was a junior associate at a relatively small investment firm, and owing to Wall Street's predominance of males, was "one of the only chicks in the room," as she told us. That didn't intimidate her, though. "It's not like I don't add value to a conversation."[1]

The host of the event was Gary Crittenden, the chief financial officer of Citigroup. At the time, Wall Street was being rocked by the unraveling of hedge funds, the meltdown of the mortgage industry, and problems with trusted investments like money market funds and previously secure firms like Bear Stearns. As the analysts chatted with Crittenden, one of the highest ranked analysts on Wall Street said he had given up trying to analyze Citigroup because it was too difficult. Hearing that, Whitney said, "I fell back on my heels." Why should it be harder to analyze the largest bank in the world than any other financial institution? And how could any self-respecting analyst intentionally avoid doing so? "That's your job. How could you have given up?" she asked.

"I thought, he may not have been able to do it, but I can do it," she said. "It kind of put a bee in my bonnet, though that's an uncool way to put it. I had a point to prove that any company is analyzable. How can you take Citigroup, such a revered institution, and say you don't understand it? You can't say they're off limits because you don't understand."

In mid-September, Whitney set about scrutinizing Citigroup's balance sheets and was astonished by what she found. Though the bank was known to have problems, the extent of those problems was much greater than anyone had acknowledged—so great, Whitney realized, that it had the potential to shake the global economy.

"I was just looking at Citigroup in the most basic, simple manner," she said, adding that as she discovered increasing evidence of financial fallacies, "I went several steps further." The bank "had gone through multi-billion-dollar acquisitions without raising capital. It was mathematically unsustainable. And all of the data was readily available."

On October 31, 2007, Whitney rang one of the biggest alarms in Wall Street history. It was just one report from one young analyst at a mid-tier research house, but Whitney's call to sell Citigroup rocked the underpinnings of the global economy. No written or spoken word has had a more powerful and immediate impact on the economy than her downgrade of Citigroup to the status of an "underperformer," the equivalent of a call to sell the stock.

Whitney was issuing quite a warning: the world's largest bank, a respected bastion of trust and stability, is weak; the whole banking system is weak. Her report contended that Citigroup was dangerously underfinanced—that the dividends paid out to investors were greater than its profits—and made the case that the dividend would likely have to be cut and that the bank would go bankrupt.

"Our thesis is simple," her report noted. "We believe over near term, Citigroup will need to raise over $30 billion in capital through either asset sales, a dividend cut, a capital raise, or a combination thereof. We believe such a catalyst will pressure the stock significantly lower." The company's stock immediately crashed, and within days its CEO, Chuck Prince, was forced to resign. But the impact went well beyond Citigroup and well beyond Wall Street.

About a month later, on November 27, 2007, Bloomberg noted that the report "hit the stock market with the force of a freight train slamming into a brick wall." Citigroup shares fell 7 percent as analysts at Morgan Stanley and Credit Suisse followed with their own downgrades. The broader market also dropped on the news, with the Standard & Poor's 500 stock index falling nearly 2.5 percent. Citigroup's board called an emergency meeting for the following weekend, and Prince stepped down three days later.

What Whitney had predicted and warned of was a disaster that stemmed from the bankers' deeply flawed thinking and actions. "This woman wasn't saying that Wall Street bankers were corrupt," wrote Michael Lewis in his book about the 2007–2008 crash, *The Big Short*, which was later made into a movie by the same name. "She was saying they were stupid. These people whose job it was to allocate capital apparently didn't even know how to manage their own." She obviously didn't cause the eventual collapse of the U.S. financial system, but she was the first to point to the dominoes just as they started to tumble.

In Lewis's view, Whitney was an oracle who came out of nowhere.

"It's never entirely clear on any given day what causes what inside the stock market," he wrote, "but it was pretty clear that, on October 31, Meredith Whitney caused the market in financial stocks to crash. By the end of the trading day, a woman whom basically no one had ever heard of, and who could have been dismissed as a nobody, had shaved 8 percent off the shares of Citigroup and $390 billion off the value of the U.S. Stock market." By some accounts, Citigroup's plummeting stock caused the bank to lose $17 billion in market value in one day. So began Whitney's odyssey from junior analyst at a comparatively obscure firm to "the woman who called Wall Street's meltdown," as *Fortune* described her in August 2008.

The warning seemed simple to Whitney, yet no one else had managed to connect the dots. Perhaps it was too hard or too risky to go up against an embattled giant. Or perhaps no one else had the ability to foresee the future impacts for the company and the broader financial world in stark, mathematical terms without fear of being disbelieved.

Some analysts, mutual fund managers, and traditional conservative stock buyers disagreed with her assessment even after her predictions began to come true—in Citigroup's case, within hours. Historically, large banks had been among the most stable, safest, and boring of investments. Among Whitney's naysayers was Legg Mason's bullish star, Bill Miller, who thought she was overreacting—a view that persisted for years, even as numerous major banks and insurance firms collapsed. Miller insisted that financial stocks would be a solid investment for the next five years because they were selling so low and, in his expectation, the banks would eventually earn their way out of trouble. Whitney, on the other hand, predicted as late as 2009 that financials had still not hit the bottom. The steady decline of Citigroup and other big banks proved her right.

WHAT DID WHITNEY SEE?

The half-decade following the turn of the millennium had been heady days for the U.S. and world economy. Despite the bursting of the dot-com bubble between 1999 and 2001, the stock market appeared to be as strong as ever. People around the world were buying houses and condos at record rates. Wall Street fueled the market with ever more exotic financial instruments, making claims of nearly risk-free returns. Today, the 2008 crash has an air of inevitability, but it shocked nearly all the stock pickers on Wall Street, a land full of prognosticators who make fortunes from what they foretell.

The primary (but not only) cause of the collapse was a crooked mortgage industry and too much of its fraudulent product bought by large banks. Wall Street investment bankers, aided and abetted by mortgage sellers, were bundling up thousands of risky mortgages in which the homeowners had a poor chance of meeting their payments (the industry calls these dogs "subprime mortgages") into aggregated pools of homeowner debt available for purchase. The bundled piles of these dangerous mortgages were called collateralized debt obligations (CDOs) and were relabeled by the ratings agencies charged with determining the likelihood they would collapse[2] as being less risky than their constituent parts.

Though the practice was widespread and more than $1.4 trillion in mortgages had been bundled into CDOs from 2004 to 2007, only a tiny handful of people realized that these CDOs were ticking time bombs placed under the largest banks, the largest businesses in the world. Warren Buffett called CDOs and other such financial products "financial weapons of mass destruction, carrying dangers that, while now latent, are potentially lethal."[3]

In the second quarter of 2007, they began to explode. As variable interest rates went up on the subprime loans that had been bundled into CDOs, foreclosures became so widespread that the mortgage industry collapsed, followed shortly by the banking and insurance industry.

Whitney saw the CDO disaster as early as 2005, and by 2007 correctly surmised that massive overinvestment by banks like Citigroup in CDOs and other risky products would lead to their downfall and cause a cascade of financial failures around the world. In fact, the following year, while the U.S. economy was in free fall, financial juggernauts like Lehman Brothers went bankrupt and the housing market—in which Citigroup was heavily invested—collapsed. The effects rippled across the world for years after.

In making her prediction, Whitney was a kind of hybrid Cassandra. She was not a classic case: she foresaw the disaster and let the world know, but the collapse may have been a foregone conclusion by the time she was heard. Whitney's predictions were initially discounted and even reviled by some, perhaps for financial reasons or because traditional conservative analysts, buyers, and money market managers simply didn't want to hear them. Who was she, anyway?

But her predictions didn't fall entirely on deaf ears, as evidenced by their impact on the stock market. Neither were her calls entirely unprecedented; some analysts and even a Citigroup senior vice president had warned of trouble weeks and in some cases years before. But no one had predicted the worst-case scenario and issued such a fact-based, specific, and inflammatory warning with potentially dire and far-reaching ramifications.

Years after the collapse began to unfold, on May 2, 2010, Whitney was still foretelling escalating doom. She told *Fortune*, "I feel like I'm at the epicenter of the worst financial crisis in history." Her original warnings had been audacious yet spot-on and had "dropped jaws from

New York to London," the magazine noted, and she had followed with prescient forecasts of more losses at Bank of America, Lehman Brothers, and UBS.

By August 2009, *Fortune* noted, most of Whitney's peers were searching for light at the end of the tunnel, but her position was that the tunnel was collapsing. In her view, bank stock investors who jumped back in would be crushed because the banks faced even worse credit losses than had so far been reported. She also predicted that the economy was about to sink into an "early 1980s-style" recession, which would devastate people and institutions that had become overextended and overleveraged during the housing boom.

As the catastrophe unfolded, Whitney's words became more than predictions; they were catalysts. Gus Scacco, an institutional fund manager for AG Asset Management, told *Fortune*, "It's gotten to the point where Meredith can't opine or write anymore without moving stocks."

There were other seers in the 2008 collapse, but not other heroes. A band of wily hedge-fund managers unraveled the bundled-mortgage fiasco and contrived a way to profit from it. Names like John Paulson, Michael Burry, and Steve Eisman have become legendary in investing circles for the billions they made and the conviction they had in what were, at the time, considered totally irrational trades. But they differed from Whitney in that they were solely in it for personal monetary gain — they sought to benefit financially from the industry's collapse. They were able to bet against the housing market precisely *because* no one believed the market was about to go bust. Whitney didn't seek to gain financially from the collapse, at least not in the most obvious way. She had a lot to gain as a financial analyst if her predictions proved accurate, but her primary aim was to warn her clients, workaday investors like pensions and mutual funds, and in that regard she behaved like a classic Cassandra.

"Cassandra" is not a static state. In the years since, her detractors have repeatedly pointed out that being right about one thing does not make a prognosticator reliable. A Cassandra is only as good as her last correct prediction. So in addition to the challenge of figuring out when someone is correctly predicting the future, there is the question of knowing when or whether to rely upon a proven Cassandra's predictions again.

In the financial world, the term *Cassandra* is typically used to describe someone who can predict rises and falls in the stock market. It is generally used as a positive descriptor despite the actual tragedy of our mythical character. Often overlooked is the fact that Cassandras are derided or disbelieved. Of course, Cassandras can only be properly labeled in hindsight, because it can be difficult at the time to know whether someone is a true visionary or more of a Chicken Little, the folktale character who believes the world is coming to an end and frequently exclaims, "The sky is falling!" Since 2007, Whitney has been labeled both.

CNBC's Jeff Cox noted on February 11, 2011, that when Whitney later overstated an impending crisis involving municipal bonds, one observer called her a Chicken Little and accused her of relying on "innuendo and assertions and hype." Another said she "spread misinformation." Among the characterizations on the opposite end of the spectrum, the June 12, 2013, *New York Observer* headlined a story "The Return of Wall Street's Cassandra: In Armani and Pearls, Meredith Whitney Smacks Back."

Most observers now give Whitney credit for being right about Citibank and the other big lenders in 2007, yet chastise her for incorrectly predicting a double-dip recession in 2011, which raises a question: Is it possible to be a Cassandra in one case and a Chicken Little in another?

A "GLAMMY" DATA-DRIVEN SKEPTIC "IN RED PATENT-LEATHER HEELS"

In the following chapter, we will attempt to identify a series of traits that can help distinguish a Cassandra. We believe there are some obvious examples and telling clues in individual patterns of behavior. Whitney echoes our other Cassandras in her dedication to the data, her consistent questioning of her analysis, and her core skepticism.

Whitney is a skeptic and loves to slice through the nonsense to get to the facts. As in the cases of so many of our Cassandras, the facts were numbers available to anyone. As she put it, "We dealt with so many sleazy managements, I never relied on management for the truth. I relied on the numbers."[4]

Meredith told us that she thinks many analysts fail to be skeptical enough, that they fall for fast talk. A common error is that we believe that "people who talk in a way you *don't* understand know exactly what they are talking about, when actually, people who talk in a way you *do* understand know exactly what they are talking about."

We also are beginning to see a pattern that Cassandras can be personally off-putting and may be eccentric or combative. Whitney doesn't look like most Cassandras in that regard.

Though she has a history of bearish views and clear talk, she is engaging, attractive, and camera-friendly, which may be one of the reasons her critics dismissed her. She was a pretty young woman "in Armani and pearls" in a field dominated by men. She doesn't have an MBA and began with no formal training in financial analysis (when R.P. first met her she was a history major at Brown University). Not to mention that she is married to a six-foot-six-inch former pro wrestler whose stage persona was a financier, and whose appearance in the ring was heralded by the sound of a Wall Street trading bell.

A *New York* magazine profile noted that Whitney "pulls off a winning high-low combination—a brainiac with an Ivy League pedigree (Brown) and a glammy party girl who's married to a WWE wrestler." These are clues as to why many people were suspicious and resistant to her calls. The biases of critics forced them to focus on her seeming overconfidence, overtly stylized manner, and gender and caused them to ignore the message, the underlying and unequivocal data.

"I'm not an old white dude, so I stick out." She took one reporter on a tour of her offices "dressed in a tightly fitted plum velvet jacket and towering red patent-leather heels," "gesturing grandly" and bragging "in her breathy voice" about having American Express CEO Ken Chenault do events there and about her plans for hosting salons. At one point, the article noted, she pivoted on her heel and proclaimed, "I'm fucking taking it on, right?"

As a young analyst, she embraced her role as a relative outlier. "As a group, sell-side stock analysts have a reputation for being reluctant to call to task the companies they cover."[5] But Whitney was skeptical of the work her colleagues were doing and even more so of the companies they were covering.

She had issued sell ratings on stocks such as Enron, Providian, CIT, and Capital One, and her buy calls were equally surprising. She had issued a buy on Goldman Sachs Group in 2006 when its shares were at $130, after which the stock soared to $200. Whitney's former colleague Steve Eisman (who profited wildly from the '08 collapse and was often guided by Whitney's research) said, "Unlike a lot of sell-side analysts, Meredith is willing to question what companies say."[6] Not being a member of the old-boys club of financial analysts and being skeptical to start with allowed Whitney to question and conclude in a way her industry simply did not before.

The combination of skeptic and "glammy party girl" didn't sit well with many people in the financial world when Whitney began issuing what seemed like rash pronouncements that had the potential to imperil financial giants. It would have been useful, on many levels, for a great many people, for her to be wrong. But she had by then taken center stage. "Who is Meredith Whitney?" columnist Thomas Friedman later asked in the *New York Times*, then noted that when he typed the first four letters of her first name into Google, it autofilled the rest.

Meredith Ann Whitney was born in 1969 in Summit, New Jersey, and graduated from the Madeira School in Virginia in 1987 before attending the Lawrenceville School for a post-graduate year and then graduating with honors from Brown University in 1992 with a BA in history. She wrote, "Studying history is about learning stories—stories about what led to seminal events and then sequels that explain what happened next and why. I love stories. My brain instinctively works to connect the dots in life, turning mosaics of information into narrative tales of how things came to be and what I think will happen as a result."[7] That analytical nature would drive her career and be highlighted by her Citigroup call.

After college, in 1993, she joined Oppenheimer and two years later segued to the company's specialty finance group. Her first job was covering what she called "one of the sketchier neighborhoods in the financial world—subprime mortgage lenders. Companies like subprime auto-finance lenders, home-equity lenders, and credit-card lenders." Most of them had no idea they were living on borrowed time, she wrote in *Fate of the States*. "Access to the market is tied to perceived ability to repay. In the late 1990s, investors started to lose faith." Many of the lenders subsequently crashed. She described the denizens of those sketchier neighborhoods of finance as "ultra-tan guys with lots of gold jewelry making subprime loans." Yet, she noted, a decade later big banks, including Citigroup, were doing the same things, and "I knew this story would end badly largely because I had lived through it once before."

Though she was not particularly well known as an analyst before she made her Citigroup call, *Forbes* magazine listed Whitney in 2007 as the second-best stock picker in the capital-market industry, behind Jeffery Harte, a principal of Sandler O'Neill & Partners who would, notably, later take a softer approach on Citigroup.

The January before Whitney's Citigroup downgrade, CNBC reported that the company had posted record revenues of $89.6 billion for the previous year, though profits were down. Prince, then still its CEO, was under pressure to boost the bank's bottom line, and some analysts said it was time to replace him. In a *Power Lunch* interview, Harte and Alan Murray of the *Wall Street Journal* told CNBC that Prince had inherited problems ranging from scandals to compliance issues, but Harte contended that the CEO had a good strategic plan and would not be going anywhere in the near future. The article noted, "There are several places investors look in bank stocks: capital markets, corporate exposure, international exposure and opportunity for improved efficiencies," adding that in Harte's view, "Citigroup meets all those criteria." As late as August 2007, Harte told *Barron's* that Citigroup could show earnings growth that it "hasn't been able to show the last couple of years" and observed that its new chief financial officer was making good decisions. He said the bank had been cutting expenses, which had been outgrowing revenues, and continuing to increase its international exposure.

The myriad dangers of a Citigroup bankruptcy may have influenced the bullish stance of some analysts, but overall, Whitney told the *Times* of London in November 2007, there seemed to be a general reluctance to rock the boat. "People are scared to be negative, especially when a company has such a wide holding," she said.

THE LARGEST COMPANY IN THE WORLD LOSES $220 BILLION, 98.5% OF ITS VALUE

Citigroup was not only the largest bank in the United States but the largest publicly traded company and bank in the world as measured by total assets, with 357,000 employees and the world's largest financial services network, spanning 140 countries with approximately sixteen thousand offices worldwide. The company had been formed in October 1998 by one of the largest mergers in history, combining Citicorp bank with the financial conglomerate Travelers Group.

Beginning in June 2006, Citigroup Senior Vice President Richard M. Bowen III, the chief underwriter of the bank's Consumer Lending Group, warned its board of directors that the extreme risks being taken on by the mortgage operation could result in massive losses. The lending group bought and sold $90 billion worth of residential mortgages annually, and Bowen said that when he first sounded the alarm, 60 percent of them were defective. The number of bad mortgages at Citigroup increased throughout 2007 and eventually exceeded 80 percent of the volume. Many were not only defective, but fraudulent, and Bowen reportedly tried to rouse the board via weekly reports and other communications.

Before Whitney's downgrade, in April 2007, Citigroup announced a restructuring to cut costs and bolster its underperforming stock that would require eliminating seventeen thousand jobs. Yet even after securities and brokerage firm Bear Stearns ran into serious trouble in the summer of 2007, Citigroup excluded its CDOs from its risk assessment, saying that their share of the bank's business was small (less than a hundredth of 1 percent). On October 15, 2007, a little more than two weeks before Whitney made her call, Bloomberg reported that Citigroup's net profit had fallen 57 percent on fixed-income losses during its third quarter and announced that mortgage delinquencies would increase and consumer lending would deteriorate for the rest of the year. Citigroup experienced its biggest share-price drop in two months after its CFO, Crittenden, said on a conference call that borrower defaults were "accelerating," according to Bloomberg.

Prince, who oversaw a 17 percent drop in the company's stock in 2007, at the time countered that momentum "continues very strong" in most of the company's businesses, though Bloomberg noted that Citigroup shares were largely unchanged since 2003 compared with a 29 percent jump at Bank of America, the second-largest U.S. bank by assets.

Despite Citigroup's troubles, Whitney's competitor, Jeffery Harte, was circumspect in his reaction. "They certainly had a lot of troubles and to some extent have been tripping over themselves the last couple of years," he told Bloomberg. He added that Prince was "doing the right things strategically. It's become more of an execution problem lately." Citigroup issued a statement saying its net income had declined to $2.38 billion, or 47 cents a share, from $5.51 billion, or $1.10, a year earlier.

It was obvious the situation was deteriorating, but so far no one had predicted the potential breadth of the problems at Citigroup and the other big banks, much less that they could undermine the world economy. Bloomberg noted that Whitney began work on her report after Citigroup reported its dramatic decline in earnings and took a third-quarter write-down of $6.5 billion, and that she had issued the report to coincide with the Fed's October 31, 2007, meeting, which warned of slowing economic growth.

"Whitney reasoned that given the current economy, the bank didn't have the means to boost its capital ratios through organic growth," the article explained. "She argued that cutting the dividend or selling assets was the only quick way to raise cash." She also said that "in six to eighteen months, Citi will look nothing like it does now.

Citi's position is precarious, and I don't use that word lightly. It has real capital issues."

The subprime-mortgage market had already fallen apart, but the bankruptcy of Lehman Brothers and the eventual paralysis of the financial system were then still unimaginable. The thought that Citigroup might become insolvent seemed absurd. Deutsche Bank analyst Michael Mayo had urged investors to dump Citigroup shares two weeks before Whitney's call, but the combination of marshaling inarguable data, predicting the extent and ramifications of the bank's problems, being sure of herself, and being media savvy transformed Whitney into the most famous and feared person in finance, New York magazine noted.

Within days of Whitney's warnings, on November 3, 2007, Bowen e-mailed Citigroup's chair and the bank's chief financial officer, head auditor, and chief risk-management officer to reiterate the risk and potential losses, requesting an outside investigation of his business unit. The subsequent investigation revealed that the Consumer Lending Group had suffered a breakdown of internal controls, yet Bowen later said his assertions were ignored despite the fact that withholding such information from shareholders violated the Sarbanes-Oxley Act (SOX). Prince signed a certification that the bank was in compliance with SOX, and Citigroup eventually stripped Bowen of most of his responsibilities and told him his physical presence was no longer required at the bank.

Bloomberg reported that the company, which declined to comment on Whitney's analysis, continued to maintain that it could rebuild its capital ratio by mid-2008 without a dividend cut or massive asset sales. "Not everyone agreed with Whitney's call on Citi," the article reported. "Jeremy Siegel, finance professor at the Wharton School of the University of Pennsylvania, said it wasn't clear that Citi would have to cut its dividend. He noted that the bank had weathered a financial crisis in the early 1990s without taking such a step." Others were harsher in their assessments. As the *New York* magazine profile put it: "Though the banking sector is melting down just as Whitney predicted it would, the financial world remains crowded with Whitney critics." The *Wall Street Journal's* David Weidner observed in April 2009, "Hers is not a one-woman show. She's had a lot of help from the media."

Criticism of Whitney continued even after her predictions came true. "Clients are not pleased with my call and I have had several death threats," she told the *Times* of London. "But it was the most straightforward call I've made in my career and I am surprised my peer analysts have been resistant. It's so straightforward, it's indisputable." She told *New York* magazine she was shocked by the intensity of the reaction. "I knew it would be a big deal, but I didn't know it would be a market-crashing event," she said. "It provoked fury amongst people. Rage, fury, and the dismissal of it, like, 'Oh, what does she know, you know, who is she?' I was just like, Whatever. If anyone ever told me that my math was wrong? That would have gotten me. That's worse than telling me that I'm fat."

"The reaction was so vicious, I was actually scared," she told us. "But ultimately, I did the right thing, because a call is a call."

In addition to death threats, she received numerous abusive e-mails and phone calls, and as a way of coping, worked out with a celebrity trainer she'd met at Bikini Boot Camp, a fitness retreat in Mexico. During a sparring match, she chanted, "I made the right call! I was right! I was right! I was right!"[8]

In the flurry of news articles about her, Whitney continued to sound the alarm about the big banks. Bloomberg quoted her saying she was fascinated by the way power is used and misused on Wall Street and attributed many of the problems in the corporate world to ego. "So much of this is a human story," she said. "Hubris is the cause of management mistakes 90 percent of the time." Since she had issued her report, Citigroup had lost $50 billion in value. "It remains to be seen whether her predictions about Citi's dividend and capital structure will be vindicated," Bloomberg noted. "But so far, in the words of one senior investment banker, 'the market is betting on her being right.'" By that point, Whitney had "emerged as a key voice of concern in this year's increasingly serious credit market crisis."

As right and strong as she was, the critics may have muzzled another critical warning from Whitney: Bear Stearns was going to collapse too. *New York* magazine noted she was "convinced that Bear Stearns was insolvent during the week leading up to the company's emergency sale to JPMorgan in March 2008. She'd been cowed by the reaction to her Citigroup report, however, and decided not to publish a piece about Bear."[9] The author of a book on the collapse of Bear Stearns quotes her as saying, "It was a conscious choice not to write anything. Because I thought it was such a tenuous situation that I was going to get in serious trouble."[10] She told us that she grew cautious after receiving "such vicious backlash from the Citi call" and having experienced "good clients turn against me." But she discounts any characterization that she softened her analysis of the companies she covered. "I put out some really powerful reports, very technical," she said. "There was nothing soft about them."

Even with the Citigroup call, she told us she had been cautious. "I sat on the report for two weeks after I printed it out," to wait for a pending announcement by the Federal Reserve Bank. "I'm always questioning myself to make sure I have the right data and then that I've done the correct analysis of the data. I beat it to death before I go live with it. It's just part of my process. Then there are industry people I check with. . . . 'Am I insane? Am I looking at this the right way?'" Though the criticism of her call was at times unsettling, "I didn't pay a lot of attention to it," she said. "There was still so much work to do. I just kept my head down and kept working."

In some ways, Whitney was more of a Cassandra *after* the Citigroup call, resisting arguments that the worst was past, that the banks could deal with the ongoing crisis. In that regard, she was ignored to a greater extent than in her original prediction about Citi's troubles.

During her research, she talked with other analysts and briefed the Federal Deposit Insurance Corporation (FDIC), which she describes as "the only organization that was prepared for subprime." She told the FDIC representatives that the banks were "naked," that the leverage rates were skewed and that all the banks were wrong. "Wachovia was the most wrong," she said. She also established a link between CDOs failing and incorrect ratings, the latter of which she argued were not just incorrect: "They were false."

By November 2008, Citigroup was insolvent despite having received a $25 billion taxpayer-funded bailout; it announced plans to cut about fifty-two thousand jobs in addition to the twenty-three thousand job cuts already made. The bank said it was unlikely to be profitable again before 2010. After that announcement, falling shares reduced the company's market capitalization to $6 billion, down from $300 billion two years prior, according to the November 23, 2008, *New York Times*. Eventually, staff cuts totaled more than a hundred thousand employees, and the largest bank in the world lost 98.5 percent of its value.

A TWO-TIME CASSANDRA?

Despite the accurate Citigroup call, Whitney's record as a seer was uneven afterward. Based on the performance of her buy and sell recommendations relative to her industry peer group, her stock picking ranked 1,205th out of 1,919 equity analysts in 2007 and 919th out of 1,917 through the first half of 2008. "That said, evaluating Whitney solely on the timing of her buys and sells misses the point," the *Fortune* article noted. "It's not just that she's bearish on the entire banking industry. What makes Whitney so interesting is the brutality of her arguments and the evidence she summons in making them." Among her more brutal arguments was that the "incestuous" relationship between the banks and the credit-rating agencies during the real-estate bubble would have a long-lasting impact on banks' ability to recover.

Whitney left Oppenheimer in February to establish her own firm, Meredith Whitney Advisory Group, or MWAG, producing company-specific equity research on financial institutions and analyzing the sector's operating environment. The same month MWAG opened its doors, Citigroup announced that the U.S. government would take a 36 percent equity stake in the company by converting $25 billion in emergency aid into common stock and would provide a U.S. Treasury credit line of $45 billion to prevent bankruptcy. In exchange, the salary of Citigroup's CEO was set at $1 per year, the highest salary of employees was limited to $500,000 in cash, and any compensation above that amount had to be paid with restricted stock that could not be sold by the employee until the government aid was repaid. The government also took control of half the seats on the board of directors, and senior management was subject to government-ordered removal if the bank's performance was deemed poor. Through all of this, Whitney—by then, a regular commentator on CNBC and Bloomberg—argued that the situation would continue to worsen through 2008, which proved true.

Whitney has not since replicated her Citigroup call and, in fact, was roundly chastised for a later bearish—and by most accounts, inaccurate—call about potential defaults on municipal bonds. Her inability to follow up with an equally momentous prediction had precedents among previous financial Cassandras. "The field tests even hard-earned reputations. Elaine Garzarelli, then an analyst at Shearson Lehman, predicted the stock-market crash of 1987 and rocketed to Wall Street stardom overnight. She launched her own business, Garzarelli Research, in 1995. If you haven't heard of her, though, there's a reason: She hasn't replicated that feat since. The pressure is on for Whitney to continue to be right and prove her value in a world saturated with expert opinion."[11]

The year 2010 was momentous both for Citibank and for Whitney. That year, Bowen testified before the Financial Crisis Inquiry Commission about Citigroup's role in the mortgage crisis, and by year's end, Citigroup had repaid the emergency aid in full, and the U.S. government received an additional $12 billion profit from selling its shares. Whitney, meanwhile, made her controversial prediction on municipal bonds, which gave ammunition to critics seeking to relegate her to the role of a Chicken Little. On December 19, 2010, in an interview on 60 Minutes, Whitney predicted "significant" municipal bond defaults, totaling "hundreds of billions" of dollars in losses, from fifty to a hundred counties, cities, and towns in the United States, saying that would be "something to worry about within the next twelve months." After the defaults failed to materialize, she described her forecast as a "guesstimate" and said some of her predictions had been misquoted.

Her detractors piled on, prompting one of her perennial champions, Michael Lewis, to reply, "Many of the articles attacking her accused her of making a very specific forecast—as many as a hundred defaults within a year!—that failed to materialize. . . . But that's not at all what she had said: Her words were being misrepresented so that her message might be more easily attacked."[12] CNBC reported on

February 11, 2011, "In the months since she made the muni call—
first on CNBC, then on '60 Minutes'—Whitney has been like a
clown sitting on the dunk tank at the county fair, just waiting for one
analyst after another to throw hardballs at the bull's-eye to drop her
into the drink." Despite that, one high-profile hedge-fund manager,
Jim Chanos, went to bat for her, saying her call was "in a general
sense" correct.

Through it all, and despite her later bullish efforts, Whitney continued to be derided as a fearmonger, as someone aligned with the
"shorts," usually hedge funds, betting that stocks would go down, to
which she replied, "That's ridiculous. I've saved long-only funds a lot
of money. It's so offensive to suggest I'm someone's puppet. I don't
identify myself with a bear-market call. I don't need a bear market to
stay in business."

Still, many depicted her as a one-trick pony, something that is
often true of Cassandras. Some critics, such as the *Wall Street Journal*'s David Weidner, even questioned her depiction as a Cassandra
in her Citigroup call. Among other voices, highly respected analyst
Dick Bove discounted such criticism, telling Weidner, "She made a
brilliant call, especially about the dividend. No one had the guts to
make that call."

Looking back, Whitney said to us, "I told a story as straightforward and simple as it could be told. Anyone could have done it. They
couldn't attack the facts, so they attacked me personally." Her goal,
she said, was "to empower the reader and empower the investor. The
math was the math." Calls such as Mayo's and warnings such as Bow-
en's were different, "Because mine was deeply quantitative. Theirs
were qualitative. For that reason, it was much more powerful, therefore irrefutable. They were subjective and qualitative."

When we asked her how she determines whether an analyst is a

Cassandra or a Chicken Little, Whitney replied, "A Chicken Little makes the same call over and over again. I make the call and move on." When someone comes to her with a prediction that's based on qualitative analysis, she said, "It's provocative. I'm open to other opinions. I've grown up feeling like the luckiest person in the room, certainly not the smartest person in the room, so I want to learn from everybody."

When she made her Citigroup call, she said, "I was with a small firm; it wasn't political; I could get away with making a call that big." She said she is well aware that some people thought she wasn't qualified to do it. "No one said it to my face, but what an amazing advantage, to be consistently underestimated," she said.

"It's not pleasant for people to look at you like you have no brain, but then you can prove to them otherwise," she said, then added, "People are mean . . . it bums me out," but said she can't afford to pay attention to all of the criticism because "that doesn't get me anywhere." The obvious question is: Why was no one else able to do it? "I kept asking myself that time and time again," she said. In her mind, "Anyone could do it . . . anyone could have."

Given the comment she overheard at the cocktail party at Citigroup's headquarters in the summer of 2007, the question needs to be asked: Were those who failed to see the extent of Citigroup's problems or rejected her call complicit, lazy, or ill informed? "Maybe all of the above," Whitney said. "They're not lazy people; they just weren't thinking about it the right way. They were smart people who were just wrong."

The data, she reiterated, was available to anyone. "People didn't want to hear it," she told us. "How dare I speak with such conviction? It wasn't me coming up with something; it was the numbers. I had conviction in the numbers; it has always been about the data and the numbers." It wasn't about sounding a fearful alarm that the sky was falling; it was about correctly foreseeing an impending disaster, based on quantifiable math.

Every piece of data Whitney used was public. "It should not have been controversial. It was not my ideas, not my own data." It just took expertise, skepticism, and being a relative outsider to see what that data foretold for the future of the financial world. And in that case, she was right.

CHAPTER 9

The Cassandra Coefficient

One man with the truth constitutes a majority.
—SAINT MAXIMUS THE CONFESSOR

The preceding chapters were stories of gloom, and made more painful by the fact that in each instance a Cassandra was pounding the table and warning us precisely about the disasters that came as promised. The people with the power to respond often put more effort into discounting the Cassandra than saving lives and resources. Thus, each of these predictions turned into Cassandra Events: warning given, warning ignored, and, of course, catastrophe.

There are many systems and techniques being used today by governments, financial advisors, investors, and futurists to look over the horizon to detect disasters. We are unaware, however, of anyone who is using a technique to seek out possible Cassandras and vetting them and their warnings against the qualities and experiences of past Cassandras. We think that such a technique should be developed and employed. In this chapter we suggest how that might be done, and in the rest of the book we apply our technique to some people and issues that today may be exhibiting the telltale signs of future Cassandra Events.

We call this, our technique of identifying predicted future disasters that are being given insufficient attention today, the Cassandra Coefficient. It is a simple series of questions derived from our observation of past Cassandra Events. It involves four *components*: (1) the **warning**, the threat, or risk in question, (2) the **decision makers** or audience who must react, (3) the predictor or possible **Cassandra**, and (4) the **critics** who disparage or reject the warning.

For each of the four components, we have several *characteristics*, which we have seen appear frequently in connection with past Cassandra Events. For example, for the issue or risk, is it technically complex? For the audience or decision maker, is there a diffusion of responsibility? For the predictor or potential Cassandra, has she been a serial false prophet? For the critics, are they arguing on the basis of data or emotion?

We are skeptical of precision in these sorts of tools for managers and decision makers. For the potential future Cassandra Events we examine in the rest of the book, we assign a score to each of the four components (high, medium, low, or absent) that indicates how well that component conforms to the pattern of past Cassandra Events. That component's score is based on our impression of how many characteristics are present and in what strength. The result is a four-column matrix. If a prediction scores all or mostly high in each of the components of this matrix, we think you had better take a closer look.

The Cassandra Coefficient does not pretend to be a complex algorithm. We expect that, for professional social scientists and those who develop high-speed stock trading algorithms, the Cassandra Coefficient may seem hopelessly naive and simplistic. Perhaps it is. Our experience with senior decision makers in governments and corporations, however, suggests to us that such leaders prefer something they can unpack, understand, and apply themselves.

Our proposition is straightforward. We have seen experts ignored in the past, when paying attention to them might have prevented or reduced the scope of calamities. In many of those cases, the same factors were at work over and over again. We can list them. If you see those things happening in that combination again, now or in the future, you may face a problem that deserves more attention and the application of a more diligent, rational, and unbiased analysis. The Coefficient can help us identify and understand our biases, the flaws in our wiring that repeatedly hinder rational thought.

While these 24 factors together are specific to finding Cassandras, many of them have broader applicability. The six factors about "The Warning" can also help us identify pending disasters that generally may escape notice. The seven "Decision Maker" factors can be indicators of bureaucracies incapable of proper awareness or response. The four factors exposing non-credible "Critics" can be relevant to generally assess detractors.

Let's take a look at the four *components* and, for each of them, the *characteristics* that we found recur in Cassandra Events. We have tried in this book to avoid social-science terminology and unnecessary complexity, but the following may get a bit jargon-intensive. Bear with us.

CASSANDRA COEFFICIENT: COMPONENTS AND CHARACTERISTICS

THE WARNING	THE DECISION MAKERS	THE CASSANDRA	THE CRITICS
Response Availability	Diffusion of Responsibility	Proven Technical Expert	Scientific Reticence
Initial Occurrence Syndrome	Agenda Inertia	Off-Putting Personality	Personal or Professional Investment
Erroneous Consensus	Complexity Mismatch	Data Driven	Non-Expert Rejection
Magnitude Overload	Ideological Response Rejection	Orthogonal Thinker	"Now is Not the Time" Fallacy
Outlandishness	Profiles in Cowardice	Questioners	
Invisible Obvious	Satisficing	Sense of Person Responsibility	
	Inability to Discern the Unusual	High Anxiety	

THE WARNING

First, what are the recurring characteristics of the issues or risks that have turned into Cassandra Events? None of the issues we have examined has exhibited all of these characteristics, but several have exhibited many of them.

* **RESPONSE AVAILABILITY:** Is this a problem that could be prevented or mitigated with some response? If not, if the

outcome is inevitable, then the issue is not one that will lead to a Cassandra Event. It may lead to a disaster, but not one in which acting on a predictor's advice would have changed the outcome. The sun around which Earth orbits will run out of fuel someday and ultimately swell to a size that engulfs Earth. There is nothing we can do to prevent that and probably nothing we can do to mitigate the results (unless you think humanity can find another planet around another star). For our purposes, we are not interested in disasters that are inevitable, or so far in the future as not to merit any meaningful response today.

* **INITIAL OCCURRENCE SYNDROME:** In many cases, the event foretold has "never happened before," at least not in the cultural memory of the audience, who will therefore resist taking the threat seriously. In our estimation, no obstacle to action is bigger than Initial Occurrence Syndrome, yet it is the easiest objection to logically assail. Doubters of relevant warnings often begin with this objection. Implicitly they are saying that nothing new ever happens, despite the manifest evidence to the contrary. History is full of examples of things happening for the first time. In fact, much of what is taught in high school history classes is simply a list of things that happened for the first time: the first flight of an aircraft, the first man on the moon, the first African American President of the United States, etc.

Social psychologists use the term "cognitive bias" to describe the filters, blinders, or limits we place between our points of view and reality. As we have seen in earlier chapters, one of the cognitive biases most relevant to our discussion is the "availability bias," a filter on perception and thinking derived from relying on familiarity or prior experience. Most predicted events that would be an initial occurrence suffer from the audience's availability bias, or lack of past experiences to which to relate the prediction.

No one in Japan remembered the high water tsunami that had taken place 1,200 years before the flooding of Fukushima; no movie or folklore memorialized it.[1] Yukinobu Okamura had to hunt for the record of the event. Rather than asking, "Has it ever happened before?" Cassandras ask, "Can it happen? Is there anything that would prevent it?"[2] If the warning is about something that has never happened in the historical record or personal experience of most people, it is likelier to be ignored.

* **ERRONEOUS CONSENSUS:** When they're first predicted, Cassandra Events don't appear as ominous dark clouds or giant tornado funnels spinning across the prairie in your direction. They require discovery and usually interpretation by an expert. The majority of relevant experts may not initially agree with the predictor who sees something new, but in times of uncertainty, the authority of expert groups can be deceiving. Physician and author Michael Crichton had this to say about the danger of consensus in science:

> *Consensus science [is] an extremely pernicious development that ought to be stopped cold in its tracks . . . it is a way to avoid debate by claiming that the matter is already settled. . . . Let's be clear, the work of science has nothing whatever to do with consensus. Consensus is the business of politics. Science, on the contrary, requires only*

*one investigator who happens to be right, which means that
he or she has results that are verifiable by reference to the
real world. . . . The greatest scientists in history are great
precisely because they broke with consensus.*[3]

In the Cassandra Events we have discussed in the
previous chapters, the majority of the "experts" agreed that
the problem was not as bad as the warning being given: the
levies would hold, the banks couldn't fail, Arab states didn't
invade other Arab states . . . The Cassandra is the outlier.
Decision makers throughout history have found comfort
in the consensus of experts and rejected or persecuted the
outlier who was often later proven correct.

Copernicus and heliocentrism, Darwin and evolution,
Hutton and deep time—the history of science is replete
with discoveries, usually by one individual, that upended
consensus. If the issue or risk defies a long-standing
consensus because of new evidence, it should be further
examined despite being a minority view.

* **MAGNITUDE OVERLOAD:** Overly large-scale events or phenomena
can have two negative effects on decision makers. First, the
sheer size of the problem sometimes overwhelms and causes
the manager to "shut down." The problem, the predicted
outcome, can sometimes simply be too big for us to get our
minds around and leads to abdication; this is called the "ostrich
effect."

Second, the decision maker may not be able to properly
magnify their feelings of dread or empathy for disasters
predicted to have massive death and loss. They simply
may not be able to grasp how enormous a threat they face.

This concept is called "scope neglect," and may best be embodied in Stalin's dictum, "The death of one man is a tragedy. The death of millions is a statistic."

Eli Yudkowsky, a possible future Cassandra we will meet later in this book, observed that:

> *Human emotions take place within an analog brain. The human brain cannot release enough neurotransmitters to feel an emotion 1000 times as strong as the grief of one funeral. A prospective risk going from 10 million deaths to 100 million deaths does not multiply by 10 to strengthen our determination to stop. It adds one more zero on paper for our eyes to glaze over, an effect so small that one must usually jump several orders of magnitude to detect the difference experimentally.*[4]

* **OUTLANDISHNESS:** In the chapters to come, we will learn about some predictions of disaster that seem better suited to science-fiction novels, Hollywood movies, or video games than to serious discussions in important rooms. Meteors, killer robots, and Andromeda strains are all ideas that have been explored in fiction and therefore seem outlandish. Instinctively, many of us feel these are the types of things "serious people" don't waste time discussing, let alone believing could be possible. As a result, due to our personal reluctance to appear amateur, we often quickly dismiss highly unconventional ideas, especially if they have only hitherto been explored on the big screen.

* **INVISIBLE OBVIOUS:** For centuries, people had watched steam escape from the teakettle, but it took James Watt to

realize this mundane daily phenomenon was a miraculous source of power that would fuel the steam engine and the Industrial Revolution. Richard Farson notes, "The most important discoveries, the greatest art, and the best management decisions come from taking a fresh look at what people take for granted or cannot see precisely because it is too obvious."[5] Farson calls this the "Invisible Obvious."

Not only can we be blinded to things because of their ubiquity or obviousness, we can also be blind to critical information because of how powerfully our attention can focus in other directions. In a psychological study in 1999, Daniel Simons and Christopher Chabris demonstrated what they called "in-attentional blindness." Because the participants in the experiment had been told to concentrate on a ball being passed from person to person, fully half of them did not notice, or at least did not react, when an individual in a gorilla suit was introduced directly into the middle of the experiment as a participant. There was nothing hiding the gorilla, the creature was obvious, but for all practical purposes it might have been invisible to half of the participants. Later Simons experimented with participants who were made aware in advance that there might be a gorilla. This time they detected the animal's presence, but failed to notice other strange things that had been introduced. Simons concluded that even when people are aware or reminded that novel or unusual things can happen, they still do not see them if they are distracted. If you are focusing on the war in Iraq, you may not notice the importance of Katrina. Cassandras see the gorilla. Cassandra Events often occur when the data is obvious if you look at it but most people are concentrating their attention elsewhere.

THE DECISION MAKERS OR AUDIENCE

The audience or decision makers are those who could do something with the warning if they believe it. In the Madoff fraud, the Fukushima disaster, the invasion of Kuwait, the *Challenger* explosion, and others, it's entirely possible the disaster could've been wholly averted if the decision maker simply "asked the next question" of the Cassandra, continued the inquiry, and wasn't so quick to dismiss her. Whether or not we simply listen to the Cassandra can mean the difference between getting pummeled by a catastrophe, or never really knowing how lucky we were to avoid one.

These decisions, as we will continue to see, are neither easy nor cost free, so we need to empower the decision makers with effective tools, tools to help recognize specifically when decision makers tend to make the wrong decisions.[6]

* **DIFFUSION OF RESPONSIBILITY:** Often it is not clear whose job it is to detect the warning, evaluate it, and decide to act. The President of the United States or the CEO of a corporation might be the person who could order action, but there may not be a general understanding of who should take the issue to them. Who owns it? Frequently, no one wants to own an issue that's about to become a disaster. This reluctance creates a "bystander effect," wherein observers of the problem feel no responsibility to act.[7] Increasingly, complex issues are multidisciplinary, making it unclear where the

responsibility lies. New complex problems or "issues on the seams" are more likely to produce ambiguity about who is in charge of dealing with them. This phenomenon is especially true when Initial Occurrence Syndrome is involved. After an event has happened for the first time, the system will decide who gets to deal with this kind of thing in the future, but if it has never happened before, it may not be readily apparent who should be evaluating the multidimensional warning.

* **AGENDA INERTIA:** Most organizations and their leadership have an agenda to which they are devotedly attached. Such groups are subject to Agenda Inertia, a force that concentrates focus on issues already in the plan.[8] Unanticipated threats, ones that the leadership didn't see coming and doesn't really want to deal with, tend to have a difficult time crowding out preconceived agenda items. Dealing with the unforeseen may take resources away from the leadership's "Pet Rock." It will certainly take away leaderships' rarest commodity, time. Warnings that have this potential to steal resources from less threatening projects tend to encounter institutional reluctance to tackle the issue. Audiences who react by rejecting an issue for these reasons tend to be the kind of decision makers who help to create Cassandra Events.

A related and equally dangerous phenomenon occurs when the organization charged with overseeing an industry becomes so vested in the industry's success that the regulators fail to criticize or properly examine the industry. This "regulatory capture" was obvious in the Fukushima disaster. The Japanese regulators were driven more by their agenda of proving to the Japanese people that nuclear power was safe than by the need to address important safety concerns.

* **COMPLEXITY MISMATCH:** Often a warning of catastrophe requires explanation or interpretation by experts and involves some technical understanding which may not be common in the audience of decision makers. When they are first predicted, Cassandra's warnings may be alarms about highly complex technologies or theories that require discovery, translation, and learning on the part of the decision maker.

The result is that some decision makers are uncomfortable with the warning, in part because of its complexity and also because their lack of expertise may highlight their own inadequacies and make them dependent upon someone whose skills they cannot easily judge.

In the case of Madoff, the SEC was outgunned when it had to consider the technical details of the supposed split-strike conversion. Harry Markopolos didn't make it any easier on them by speaking to lawyers in mathematical jargon. In this instance, like many others, the decision makers simply gave up on even trying to understand the technical complexities and ignored the Cassandra.

Systems can be so complex that even experts can't see the disaster looming within. Complexity mismatch is a looming threat for government. For the first time, technologists are now building machines that make decisions with rationale that even the creators don't fully understand. The accelerating growth of technology makes it increasingly

difficult for scientists, let alone bureaucrats, to decipher
the risks. Many new technologies have proven the limits
of government regulators' ability to understand: genetic
engineering, artificial intelligence, high-frequency trading,
deep water drilling, and others.

Mathematicians, system engineers, and social scientists
have written extensively about complexity theories for
decades. In recent years, however, we have seen a particular
aspect of complexity emerging more clearly in the real
world. Increasingly, we are operating or planning systems,
software, or networks that no one person understands.
It takes a team, one of many diverse talents. That team,
however, is sometimes so large that it cannot be assembled
in a conference room, auditorium, or even in a stadium.

In 2014, hackers realized that for two years a near
ubiquitous component of software, OpenSSL, had a
vulnerability due to the absence of a length validation.
Many corporations were not concerned with this news
because they didn't use open-source software, only
proprietary code they developed themselves. These
corporations were shocked when the Heartbleed virus
attacked their systems and stole data using the OpenSSL
vulnerability. Upon examination, chief information officers
realized that their code writers had simply borrowed
OpenSSL and incorporated it into their software. If an issue
is so complex that no one person fully understands it, there
may be an increased likelihood of a Cassandra Event.

* **IDEOLOGICAL RESPONSE REJECTION:** Some decision makers
reject the only available responses to a risk because those

paths don't conform to their ideology, for example about
the proper role of government or science. Sometimes the
ideology is an interpretation of a religion or its tenets.
To avoid or mitigate a disaster, a response may require
increased government spending, a new or larger government
organization, additional legal or regulatory constraints on an
industry, or on perceived or desired freedoms. If believing
in the risk leads inevitably to an ideologically repugnant
response, some decision makers will reject the warning
altogether.

* **PROFILES IN COWARDICE:** It takes personal courage for
leaders to detect and evaluate a warning, determine that
a disaster is coming, and order resources expended and
lives disrupted to deal with the risk. They will weigh the
personal consequences of being wrong, of playing the fool,
of becoming the public embodiment of Chicken Little. For
it is not only the technical expert, the warner, who will be
ridiculed and professionally damaged if the disaster does not
come. It is perhaps even more so the leader who was duped
into believing the warner. If the individual or collective
leadership at the top is overly cautious or lacks creativity, it
will likely turn a deaf ear to Cassandra.[9]

* **SATISFICING:** Decision makers faced with a warning of
disaster often do respond in some way, an insufficient way.
Faced with the risk that Cassandra is correct, but not really
understanding the technical aspects of the issue sufficiently to
make a personal judgement, decision makers will often make
a token response.[10] The sociologist Herbert Simon coined
the term "satisficing" in 1947 to describe searching through
the available remedies until an acceptable alternative is

found that "addresses" the problem, but doesn't solve it. This alternative is usually easy, not requiring significant resources or disruption. Ordering additional studies is a typical satisficing strategy, but may be appropriate if there is time and if the analyses truly are incomplete or insufficiently tested. Although hedging may be a rational response in some cases, when decision makers engage in a response that can't solve the problem in time or actually change the outcome, there is often a Cassandra Event in the offing.

* **INABILITY TO DISCERN THE UNUSUAL:** Especially with first-time events, but also more generally, warnings can be blocked from reaching the appropriate decision makers by the nature of the institution involved. Big organizations, whether government agencies, large corporations, or professional scientific establishments, frequently are poor at discerning the urgency of an issue. They often cannot tell the difference between the routine and the dramatic. Information that is not routine and should be rapidly escalated to higher-level decision makers is often instead placed in the queue for consideration in due course by systems not designed for determining whether an alarm is urgently required.

   Our former colleagues in the White House Situation Room faced this problem in the middle of the night several times a week. When news came in, they had to ask themselves if it was big enough to wake the President or if it should just be put in the message queue for the appropriate desk officer to deal with the next day. The Sit Room teams usually got it right, but that was because they recognized that warning was part of their job description.

The National Weather Service also issues warnings. It has sensors that look for certain classes of extreme conditions. Its personnel have protocols for what to do when out-of-the-ordinary readings come in, thus the Weather Service can act quickly to save lives. For some issues, however, the relevant bureaucratic entity has no sensors, no rules on what to do with unusual readings, no automatic procedures, no channels for disseminating alerts. Therefore, when examining the nature of the issue, we need to ask whether there really is a system with a priori–derived criteria for determining whether something is dangerously unusual.

We discuss more about this in the conclusion, but it is worth noting here that successfully finding Cassandras is unlikely to happen unless relevant organizations, both in government and the private sector, foster a culture poised to listen to warnings, and empower and train small cadres to recognize and test sentinel intelligence.

THE PREDICTOR OR POSSIBLE CASSANDRA

The individual giving the warning is, of the four components, the most important and perhaps the easiest to judge. He or she often has certain strengths and weaknesses.

* **PROVEN TECHNICAL EXPERT:** The Cassandras we have seen are recognized in their fields as competent experts. They didn't see fiery bushes while wandering alone on mountains. They weren't awakened in the night with divine revelations. They were accredited experts who, while doing their jobs and pursuing research in their areas of specialty, discovered or

were made aware of disturbing information that led them to a conclusion that others hadn't yet reached or didn't think warranted breaking the glass and pulling the alarm.

Cassandras see something no one else sees, not unlike renowned visionaries in technology who envision things that no one else has yet built. Formal education or training may be essential for their work. As professor of business psychology Tomas Chamorro-Premuzic noted, "Contrary to popular belief, most successful innovators are not dropout geniuses, but well-trained experts in their field. Without expertise, it is hard to distinguish between relevant and irrelevant information; between noise and signals."[11] We'd say the same about potential Cassandras.

Usually Cassandras have not given a dire warning before, or if they have, they were clearly proven to be right. The Cassandras that we have selected are not people who issue so many warnings that they just happen to get one right.

* **OFF-PUTTING PERSONALITY:** Because of the frustration of seeing a threat and wondering why others can't, because of their personal sense of responsibility for promoting understanding and action on their discovery, and perhaps because of their high level of anxiety in general, Cassandras may at times appear obsessive and even socially abrasive. While they can be personally charming under the right circumstances, many of the individuals gifted with the intelligence and strength of personality required to be a Cassandra may sometimes seem aloof, condescending, socially maladapted, or absent-minded.[12] Many Cassandras might score low on what is sometimes called EQ, or the emotional quotient of personal interaction skills. That characteristic may prevent

them from communicating in a way that will elicit the appropriate response to get their warnings taken seriously. The work of Lee Ross, a Stanford sociologist, demonstrates that most people are unable to differentiate a message from the messenger.

* **DATA DRIVEN:** Cassandras' warnings are not generated on the basis of intuition or "analyst's judgment." They are driven to their conclusions by empirical evidence. Often they are the first ones to generate or discover the data, but the evidence is usually not in question. It is their interpretation of the data that makes them step away from the previous consensus. They tend to see the problem leaping out of their data with a clarity that makes them unique. Some have an almost savant-like quality, an ability of instant pattern recognition in the clutter of data.

* **ORTHOGONAL THINKER:** Cassandras tend to be among the first to think about a certain problem or issue and often are those who acquire the data that then causes alarm. Because of their originality, Cassandras come at the issue from a new perspective and incorporate data and concepts from other fields. This characteristic is called orthogonal thinking. They have the self-confidence to be first but not the arrogance that would interfere with their understanding of the nuances of the data.

* **QUESTIONERS:** Most Cassandras tend to disbelieve anything that has not been empirically derived and repeatedly tested. They also tend to doubt their *own* work initially, especially when it predicts disaster. This characteristic is more than just a belief in the scientific method. Rather, they challenge what

is generally accepted until it is proven to their satisfaction. They are the philosophical descendants of Pyrrho of Elis, a philosopher in ancient Greece who accompanied Alexander the Great to India. There Pyrrho learned from Indian philosophers who challenged everything. Pyrrho's teachings influenced another Greek philosopher who taught that all beliefs and assumptions should be challenged, that doubt, skepticism, and disbelief are healthy. This later philosopher was Sextus Empiricus, and his name is forever attached in our minds to the empirical method: doubt until proven by data, by objectively true, observable facts.

Many Cassandras seem to have incorporated Albert Einstein's belief that "unthinking respect for authority is the greatest enemy of truth." When the authority figures to whom they report their warning reject their analysis for what the Cassandras believe are non-evidence-based reasons, our warners begin to lose respect for the decision makers. They often are unable to hide that disrespect well.

* **SENSE OF PERSONAL RESPONSIBILITY:** When Cassandras discover data indicating an impending disaster, they report it, usually via the channels with which they're most familiar. Typically, when that results in little or insufficient reaction, they don't easily move on to another issue or project. They feel a sense of personal responsibility to fully and clearly explain the significance of their discovery and the consequences of inaction. They tend not to understand the inability of others to see the import of what is so obvious to them. Thus they are usually driven to act to gain attention to their issue. Cassandra is the person in the crowd who smells the smoke first and is sufficiently confident in her judgment and so

filled with a sense of personal responsibility that she is the
first to call 911 or pull the fire alarm.

* **HIGH ANXIETY:** In 1977 Mel Brooks produced a comedy movie
named *High Anxiety*, about the problems a California
psychologist has with a zany crew of patients and coworkers.
Actual high anxiety, however, may correlate with the ability
to foresee disaster. In 2012, psychiatrist Jeremy Coplan, of
SUNY Downstate Medical Center, found that generalized
anxiety disorder was associated with higher intelligence
levels and was not necessarily correlated with neuroses or
dysfunctional behavior.[13] In Israel, Dr. Tsachi Ein-Dor drew
a similar conclusion, that people with higher anxiety levels
tended to detect threats sooner and warn others.[14] Such a
result might be expected in Israel, given the near-constant
threat of terrorism, but in Canada, Professor Alexander
Penney conducted experiments from which he concluded
that people with higher anxiety levels were better able to
discern and "laser focus" upon a primary threat, despite
being bombarded with secondary problems.

THE CRITICS

In addition to scoring the issue, audience, and Cassandra on the characteristics discussed above, it may also be useful to look at who the
Cassandra's critics are, what their motives appear to be, and what the
nature of their criticisms are.

* **SCIENTIFIC RETICENCE:** For some issues, a high scientific
standard of proof cannot be met in time to act. Some events
cannot be accurately created, simulated, and repeated in

the laboratory. To avoid the disaster, it may be necessary to abandon the normal protocol of waiting for all the evidence to be in and act instead on incomplete data and early indications. Scientists and decision makers in denial may argue to wait for the final results, for additional studies. Sometimes, though, waiting can prove fatal. Of course, acting with insufficient basis may also produce a disaster of a different kind. When the issue or risk under consideration involves that kind of trade-off, it may require innovative evaluative methodologies.[15]

As we are considering the world of experts and technocrats, it is worth noting that Scientific Reticence and some other errors described in the Cassandra Coefficient can actually increase as people become more learned. Luke Muehlhauser writes about the "sophistication effect," in which "the most knowledgeable people, because they possess greater ammunition with which to shoot down facts and arguments incongruent with their own position, are actually more prone to several of these biases."[16]

* **PERSONAL OR PROFESSIONAL INVESTMENT:** People with something to lose from the revelation of a risk, or from the solutions, may criticize Cassandras for illegitimate reasons based on this self-interest. An obvious example would be a member of an industry who criticizes a warner for a suggested solution that could increase costs for operating in that industry (the extractives industries and global warming come to mind). Other experts in a field may instinctively react badly to the work of a Cassandra if the new investigation unseats something that they had earlier developed or championed themselves. Some experts react negatively out of simple

jealousy that another expert, or worse, an outsider, got there first or is getting a lot of publicity. Professional colleagues may think a Cassandra is "grandstanding" or seeking public attention when she goes public with the warning, bringing it to the attention of an audience beyond the often insular professional community. Criticism focusing on these kinds of issues, rather than the data, suggests a Cassandra Event.

* **NON-EXPERT REJECTION:** Some ideas, once exposed to peer review, gain general expert acceptance but are still heavily assailed by non-experts, many of whom may have a vested interest either in keeping things as they are or otherwise not acting to prevent or mitigate the disaster. If the vocal critics are not experts in the field and don't argue on the basis of relevant data, there is cause for concern. Likewise, in the case of critics later found wrong, we have found a trend of intellectually inconsistent criticism. Individual critics may sound convincing, but it's worth looking at the body of critics to see if there is any harmony in their arguments. If different critics have different arguments for why a warning is not legitimate, especially if they're not data-driven, take a closer look at the warning.

* **"NOW IS NOT THE TIME" FALLACY:** Some opponents to acting on the basis of the evidence of an impending disaster know better than to deride the experts or question their technical data and analyses. Instead, they seek to minimize the urgency or defer consideration to some vaguely defined future. They argue that there are other more urgent issues, that resources are currently constrained, and that others should become involved in the solution before action is taken. These responses can best be thought of as "Maybe, but . . ."

MAKING THE SAME MISTAKES

The ignoring of Cassandras, like many catastrophes themselves, often results from a cascade of errors. Within that cascade, there are, as we have just reviewed, a variety of reasons why Cassandras are ignored.

History teaches us that decision makers make exactly the wrong decision about Cassandras because of mental strategies we all employ subconsciously. The eighty-six billion or so cells that are your brain were designed long ago. The brain is a very old tool, but an amazingly efficient one. It understands the value of shortcuts and simplifying strategies. Social scientists call these shortcuts "heuristics." Heuristics are the neural equivalent of video-game cheat codes. They let us skip through levels of slow, rational decision making to rapidly decide what to do. These information-processing shortcuts are based on the recognition of patterns collected throughout our lives to allow our brains to swiftly make decisions, usually with high (sometimes unfounded) confidence. We seem to know almost at a genetic level that when we see a snake, it's a good idea to move away from it quickly. We don't need to stop and think about it.

The problem with heuristics is that they tend to become cognitive biases, inaccurate prejudices against one decision or one person versus the alternative. One form of bias is when our ancient brain gives preference to subjective points of view over rational and objective facts. Some of the characteristics we discussed above are examples of these mental shortcuts.

Many of our Cassandras were ignored because they weren't considered personally compelling by decision makers. There was something "wrong" with them, and this subjective judgment biased decision makers against looking more closely at the Cassandras' evidence. The analysts at the SEC made a biased snap judgment about Harry Markopolos. Based on the fervor of Markopolos's argument, his paranoid demeanor, and the fact that he worked for a Madoff competitor, he was immediately assessed by SEC personnel as self-interested, slightly unhinged, and ultimately not trustworthy. They ignored his well-evidenced complaint, in part because of a subjective personal bias that blinded them to the objective, data-driven proof that Bernie Madoff had stolen billions of dollars. Respect for authority is another heuristic. Trust in hierarchy can interfere with accurate decision making and become a bias.[17] In the Bernie Madoff example, it prevented the junior employees at the SEC from thoroughly investigating a highly respected denizen of the business community.

Just as decision makers often have a hard time navigating the sometimes off-putting personalities of Cassandras, the Cassandras often have problems navigating the biases of others. Rapidly judging other people is the primary skill for which our prehistoric brains are optimized, but this skill may be a dangerous barrier injecting subjective bias where we'd be better served by objective rationality. We reflexively consider the witness before we weigh the testimony. Traditional investors on Wall Street lost billions when they were unable to move past their biases about Meredith Whitney based on her youth, gender, or manner. They were also blinded by their biases that large banks were paragons of stability. These biases meant they did not see her compelling evidence-based exposition that Citigroup was teetering.

These cognitive biases are so much a part of who we are that many of the most successful hedge-fund managers realize that such mental filters can cost them hundreds of millions of dollars in trading errors. Many of them study cognitive biases to identify such flaws in their own behavior, hiring psychologists to assess them, as well as their trades, for decisions influenced by biases that occlude rational, factual analysis. MI5, the UK equivalent of the FBI, has followed a similar practice. During periods of intense stress that increase cognitive overload, such as terrorist attacks and hostage crises, MI5 assigns a psychologist to the command team to warn of incipient bias-based decision-making errors.

Even with psychologists looking over our collective shoulder, we cannot completely escape the biases of the human brain, which had its last evolutionary upgrade seventy thousand years ago.[18] However, we can try to minimize the harm they cause us if we put systems in place to deal with known, common sources of decision-making error, such as ignoring potential Cassandras.

FALSE CASSANDRAS

In addition to unconscious biases, a frequent reason decision makers ignore Cassandras is fear of being embarrassed by embracing some-one who turns out to be wrong. No one wants to be the person who believed the "boy who cried wolf," and especially not the one who spent a lot of money on that belief. Therefore, we have to have a way of filtering out false Cassandras.

There are, of course, innumerable examples of people crying out that the sky is falling or the end is nigh. These people are easily detected. They are not experts in any empirically based field. Their warnings are not based on observable data, and they tend to have a track record of other beliefs and predictions that are demonstrably false. The real problem comes when the potential Cassandra is an established technical expert who is making the case with data. There are numerous examples of people who looked like true Cassandras but were simply wrong. Some such non-catastrophes are ambigu-ous, in part because the warnings elicited hedging responses that may have affected the outcome. Looking at three of these cases may help elucidate the problem.

* **THE POPULATION BOMB**: Stanford biologist Paul Ehrlich came
  to the public's attention in 1968 when he published and
  vigorously promoted his book *The Population Bomb*. In it,
  he predicted that the growth of world population, which
  would repeatedly double in ever shortening cycles, would
  result in widespread fatal famines beginning in the 1970s
  and 1980s. He was a scientist at a prestigious school. Much
  of the demographic and other data he used was created by
  UN agencies and was widely accepted. He advocated zero
  population growth (ZPG) attained by intrusive, government-
  sponsored birth-control campaigns.

     There were few massive famines in the 1970s and 1980s,
  or indeed since, caused by a lack of food-growing capacity
  relative to global population. Indeed, on a global basis,
  caloric intake significantly increased. Localized famines
  did occur, but usually because of climate change, sustained
  weather anomalies, or wars. People did die, but usually
  because governments failed to deliver food assistance. The
  global population did increase significantly, but so did
  humanity's ability to feed itself.

     Ehrlich underestimated the ability of agronomists and
  others to increase the world's population-carrying capacity
  by improving crop yields and distribution systems. Some
  demographers failed to see that population growth rates
  would be slowed in countries by improving economic and
  social conditions, leading to ZPG or lower rates of growth
  in some highly industrialized, technologically advanced
  societies.

     Ehrlich was not an expert in demography or agronomy.
  His analytical failures seem mainly attributable to not
  taking into account feedback loops, i.e. not considering

the role that could be played by elements of the system adjusting to address the problem. As demand for food increased, research organizations like the Rockefeller Institute, Manila's International Rice Research Institute, big agribusinesses like ADM and Monsanto, and Nobel Peace Prize–winning biologist Norman Borlaug developed ways to substantially increase crop yields. Governments, international organizations, and NGOs did promote birth control. Increased levels of education and entry into the workforce of women also reduced birth rates.

There are those who believe that Ehrlich was not wrong, just premature. They think that what he predicted will happen, just not yet. In any event, one clear lesson from the population-bomb case is that decision makers need to model how the overall system will adjust when it perceives the threat and whether that will delay the disaster or significantly reduce the problem.

* **Y2K:** In 1984, Jerome and Marilyn Murray published a book, *Computers in Crisis*, which predicted that when 1999 rolled over into the year 2000, many software programs would malfunction. Although it initially gained little public attention, by the mid-1990s, software developers and computer scientists in general were in widespread agreement that there was a potentially serious problem. Many software programs could not display the year 2000 because they were written with only two digits available for the designation of the year. The year 2000 would be displayed as simply 00, which would cause some software to freeze up or go into a loop. Even in the late 1990s, many critical functions were run by software. Having software

programs associated with these functions stop on New Year's Eve could cause important financial, transportation, energy, and weapons systems to fail. It was, in many ways, the first crisis of the so-called Information Age.

Initially, it was difficult to gain the attention of senior government managers for the importance of a "software glitch." In the White House, however, a senior official in the Office of Management and Budget, John Koskinen, grasped the importance of the issue. (It is hard for Dick to think that this is an example of a prediction that was proven entirely wrong because he worked closely with Koskinen on the issue.) Koskinen, a professional public administration troubleshooter, persuaded a junior senator, Robert Bennett (R-NV) to champion an emergency program to fix the government's computers and mobilize the private sector in critical industries to do the same. Koskinen used the United Nations and the European Union to spread the message. Untold billions of dollars were spent to rewrite software in the two years leading up to December 31, 1999.

Some systems around the world did fail at the stroke of midnight. There were no major problems, however, and nothing approaching a disaster. Was it a problem averted by Herculean effort, or a disaster prediction greatly exaggerated? It was a little of both, but sometimes you need to indulge in some hyperbole to get the attention and action required.

In truth, despite modeling and simulation attempts and experiments, it was difficult to know in advance how bad the problem would be. There were no databases to accurately tell decision makers what systems were reliant upon what software and which ones were likely to fail.[19] Koskinen did not take a hedging strategy of spending a little to solve some of the problem. He saw it as an issue whose solution required significant resources, an issue for which there was no partial solution, in which some of our critical systems would be allowed to fail. The downside of his being wrong, if that happened, was quite limited. The worst thing that would happen if he overreacted was that a lot of software would be updated and many software and service companies would be enriched.

The lesson of the Y2K case study is that, if the Cassandra is right and averts the problem by action, there will be those who decry overreaction. One cannot always prove such critics wrong, but experiments and models can be used to demonstrate what might have happened in the absence of the response. If you decide to act in a big way to solve the alleged problem, it is, of course, less risky if the costs of your being wrong are limited and are clearly outweighed by the potential impact of not acting or doing so insufficiently.

* **SWINE FLU:** In January 1976, thirteen U.S. Army personnel died of a new strain of flu. Analysis of the flu determined that it was related to the great Spanish flu epidemic of 1918, which killed millions in North America and Europe, including five hundred thousand people in the United States. Experts from the Center for Disease Control and the National Institutes of Health determined that there was a possibility that the flu would reemerge in the next winter and kill millions. They were possible Cassandras, experts with data, warning of an impending disaster. Their warnings were well received in the appropriate federal ministry, the Department of Health, Education, and Welfare. Outside

experts validated their fears and their proposed solution, an unprecedented crash program to develop a vaccine and administer it to "everyone" in the United States by the end of the year. President Gerald Ford, who had succeeded to the presidency because both the Vice President and the President had been forced from office, accepted the recommendations and asked Congress for swift approval. Congress gave it.[20]

When insurance companies balked at issuing policies for the vaccine makers, Ford successfully asked Congress for an unprecedented liability waiver. When vaccinations commenced, there were problems. The dose proved too strong for children and there was no pediatric dose available. There were reports of neurological disorders in a handful of vaccine recipients. In many places, the program was put on hold. In others, fewer people showed up to be vaccinated than had been anticipated. The country was still at risk if a deadly pandemic emerged, but it did not. Indeed, the swine flu never emerged at all. The virologists had erred in their modeling of this flu.

In their decision-making process, the federal officials had explicitly said that the downsides to acting were few and the potential costs of inaction were millions of dead Americans. Although the scientific and public-policy analyses involved in the decision were flawed, as amply demonstrated by Harvard historian Richard Neustadt (*The Epidemic That Never Was*), the issue, framed as it was, gave President Ford little choice. There were, however, some costs because the threat never materialized. Some people were sickened by the vaccine. The credibility of the public health system was damaged, and that may have resulted in some people refusing to get necessary vaccines in the future (although there is little evidence that has happened in any significant way). Ford seemed in over his head, perhaps contributing to his loss in the election in November of that year, 1976.

Neustadt's major criticism of the government reaction to the Cassandras could be said to be the inverse of Initial Occurrence Syndrome. What drove the decision, he claims, is that something like the predicted pandemic killer flu had actually happened before (in 1918) and the death of nearly a hundred million people was in the folk memory of decision makers. The "availability bias" in this case *supported* the fears of the Chicken Littles.[21]

Neustadt took other lessons from his case study, regarding the interaction between the technical expert community and the policy community. There was not one warner, but rather the assembled virology and public health experts used by the federal government.[22] Neustadt suggests that there were doubts in the expert community, but they were never made explicit to, or discovered by, the policy community. The expert community had a vested interest in strengthening the public health and vaccine system. Neustadt believes that a good policy-analysis effort would have reduced the estimate of the health risk and would have identified numerous implementation risks.

A hedging strategy was not fully explored and might have been available. The vaccine might have been stockpiled throughout the country and delivered to the population on a crash timeline if the flu emerged. There were risks in that too. Some people would likely have died in the early wave of the flu. Nonetheless, a surveillance or overwatch strategy might have been available in this instance, and with better policy analysis, exposing analytical assumptions more explicitly, it might have been taken.

Let us continue by looking at seven more case studies. In the subsequent chapters we will try to apply the lessons we have learned and the Cassandra Coefficient to people among us today who are predicting major disasters, but are not being heeded. Should they be? Are they future Cassandras? Let's listen to them and find out.

CHAPTER 10

The Computer Scientist: Artificial Intelligence

The real problem of humanity . . . we have Paleolithic emotions; medieval institutions; and godlike technology.

—E. O. WILSON

Twenty of the wing-shaped drones launched from the deck of the supercarrier U.S.S. Gerald R. Ford *and quickly disappeared over the horizon toward the Chinese-held islands. Communications with the drones were lost quickly due to the Chinese military's jamming the command and control frequencies. Instantly, the drones switched into autonomous mode. They communicated with each other using lasers that could not be jammed. Soon, the swarm of drones, acting as one integrated attack force, spotted the ships. Matching their multispectral collections with on-board data sets, they identified the assorted destroyers, frigates, and one aircraft carrier as enemy combatants. Almost instantly, the drones ran millions of calculations and decided which of the ships was the biggest threat and would be attacked first. Then they decided which drone would attack what target. Then the assault began, with the swarming drones launching missiles at the Chinese Navy squadron from seemingly every direction and altitude.*

THE U.S. NAVY HAD THAT SCENARIO IN MIND WHEN IT DEVELOPED THE X-47B, the first drone to be launched from and recovered on an aircraft carrier.[1] The Northrop Grumman aircraft was designed to fly in swarms, identify enemy targets, and attack them autonomously, without a human in the loop.

The U.S. Air Force has also been planning for lethal autonomous drones, writing in its "Unmanned Aircraft Systems Flight Plan 2009–2047," that it was only a matter of time before drone aircraft would be making life-and-death decisions on their own using artificial intelligence (AI) software.

Those U.S. military plans for autonomous lethal systems using AI came to a halt, at least for now, when then secretary of defense Leon Panetta issued Department of Defense Directive 3000.09 in November 2012. That Pentagon policy banned the use of lethal autonomous systems. The directive was needed because the technology to create such "killer robots" had already been achieved. Indeed, ten years ago the Department of Defense had developed the LOCASS, the Low Cost Autonomous Attack System, a jet-powered bomb that would fly around until it saw what it thought was an enemy vehicle and then attack it, again without a human in the loop. While technology exists for AI-driven weapons, we humans are not quite up to speed on how safe they are or will be, or how to use them.

MACHINES THAT LEARN

Artificial intelligence is a broad term, maybe overly broad. It simply means a computer program that can perform a task that normally must be done by a human. Such tasks include decision making, language translation, and data analysis. When most people think of AI, they are really thinking of what computer scientists call weak artificial intelligence,[2] the type of AI that runs everyday devices like computers, smartphones, even cars. It is any computer program that can analyze various inputs, then select and execute from a set of preprogrammed responses. Today, weak AI does relatively simple (or "narrow") tasks: commanding robots to stack boxes, trading stocks autonomously, setting the air-fuel mixture in car engines, or running smartphones' voice-command interfaces.

Machine learning is a type of computer programming that helps make AI possible. Machine-learning programs "have the ability to learn without being explicitly programmed,"[3] optimizing themselves to most efficiently meet a set of pre-established goals. Machine learning is still in its infancy, but as it matures, this ability to learn, for computers to reprogram and improve themselves, makes the future of AI dramatically different from any invention that has come before.

Computers teaching themselves, making themselves smarter, leads us to "superintelligence."[4] Superintelligence is an artificial intelligence that will be "smarter" than its human creators in all the metrics we define as intelligence. Superintelligence does not yet exist, but when it does, some believe it could solve every major problem of humankind: aging and disease, energy and food shortages, climate change. Self-perpetuating and untiring, this type of AI will improve at a remarkably fast rate, and eventually surpass the level of complexity humans can understand. This is the promise of AI, and possibly its peril.

Experts have given a name to this era of the hyperintelligent computer: the "intelligence explosion." Nearly every computer and neural scientist with expertise in the field believes that the intelligence explosion will happen in the next seventy years; most predict it will happen by 2040. In 2015, more than $8.5 billion was invested in the development of new AI technologies. IBM's Watson supercomputer is hard at work performing tasks ranging from playing (and winning at) *Jeopardy!* to diagnosing cancer. What will Earth be like when humans are no longer the most intelligent things on the planet? As science fiction writer and computer scientist Vernor Vinge wrote, "The best answer to the question, 'Will computers ever be as smart as humans?' is probably 'Yes, but only briefly.'"[5]

As the excitement grows, so too does fear. The astrophysicist and Nobel laureate Dr. Stephen Hawking warns that AI is "likely to be either the best or worst thing ever to happen to humanity, so there's huge value in getting it right." Hawking is not alone in his concern about superintelligence. Icons of the tech revolution, including former Microsoft chairman Bill Gates, Amazon founder Jeff Bezos, and Tesla and SpaceX CEO Elon Musk, echo his concern. And it terrifies Eliezer Yudkowsky.

Eliezer has dedicated his life to preventing artificial intelligence from destroying humankind. Tall with a thick, dark beard that, along with wire-rim glasses, forms a frame around his large, oval face, he is a thirty-seven-year-old autodidact who dropped out of school after eighth grade. Married without children, Eliezer grew up in Chicago and now lives and works in Berkeley, California, at an organization he founded, the Machine Intelligence Research Institute (MIRI). His work focuses on foundational mathematical research to ensure (he hopes) that artificial intelligence ultimately has only a positive impact on humanity. The ultimate problem: how to keep humanity from losing control of a machine of its own creation, to prevent artificial intelligence from becoming, in the words of James Barrat in the title of his 2013 book, *Our Final Invention.*[6]

A divisive figure, Yudkowsky is well known in academic circles and the Silicon Valley scene as the coiner of the term "friendly AI." His thesis is simple, though his solution is not: if we are to have any hope against superintelligence, we need to code it properly from the beginning. The answer, Eliezer believes, is one of morality. AI must be programmed with a set of ethical codes that align with humanity's. Though it is his life's only work, Yudkowsky is pretty sure he will fail. Humanity, he tells us, is likely doomed.

SUMMONING THE DEMON

Humans will relentlessly pursue the creation of superintelligence be-cause it holds unimaginable promise to transform the world. If, or when, it is born, many believe it will rapidly become more and more capable, able to tackle and solve the most advanced or perplexing challenges scientists can propose, even those that they can't yet even conceive of. A superintelligent computer will recursively self-improve to levels of intelligence that we may not even be able to comprehend, and no one knows whether this self-improvement will happen over a long period of time or within the first second of being turned on. It will carve new paths of discovery in fields yet undiscovered, fueled by perpetual improvements to its own source code and the creation of new robotic tools.

Artificial intelligence has the potential to be dramatically more powerful than any previous scientific advance. Superintelligence, ac-cording to Nick Bostrom at Oxford, is "not just another technology, another tool that will add incrementally to human capabilities." It is, he says, "radically different," and it "may be the last invention humans ever need to make."[7]

From Yudkowsky's whimsical graph above, we can get a hint of the predicted power of superintelligence (here called "recursively self-improved AI"). Nearly all AI theorists predict superintelligence is coming and that it will so vastly eclipse the cognitive capacity of even the smartest humans that we simply cannot comprehend its intellectual horsepower. It's a lot to swallow.

Yudkowsky and others concerned about superintelligence think in terms maybe best described as Darwinian. Once humans are no longer the most intelligent species on the planet, humankind will survive only at the whim of whatever is. He fears that such superintelligent software would exploit the Internet, seizing control of anything connected to it, like electrical infrastructure, telecommunications systems, manufacturing plants . . . Its first order of business may be to covertly replicate itself on many other servers all over the globe as a measure of redundancy. In could build machines and robots, or even secretly influence the decisions of ordinary people in pursuit of its own goals. Humanity and its welfare may be of little interest to an entity so profoundly smarter.

Elon Musk calls creating artificial intelligence "summoning the demon" and thinks it's humanity's "biggest existential threat."[8] When we asked Eliezer what was at stake, his answer was simple: everything. Superintelligence gone wrong is a species-level threat, a human extinction event.

Humans are neither the fastest nor the strongest creatures on the planet but dominate for one reason: humans are the smartest. How might the balance of power shift if AI becomes superintelligence?

Yudkowsky told us, "By the time it's starting to look like [an AI system] might be smarter than you, the stuff that is *way* smarter than you is not very far away." He believes "this is crunch time for the whole human species, and not just for us but for the [future] intergalactic civilization whose existence depends on us. This is the hour before the final exam and we're trying to get as much studying done as possible."[9] It is not personal. "The AI does not hate you, nor does it love you, but you are made out of atoms which it can use for something else."[10]

Self-aware computers and killer robots are nothing new to the big screen, but some believe the intelligence explosion will be far worse than anything Hollywood has imagined. In a 2011 interview on NPR, AI programmer Keefe Roedersheimer discussed *The Terminator* and the follow-up series, which pits the superintelligent Skynet computer system against humanity.[11]

MR. ROEDERSHEIMER: The Terminator is [an example of an] AI that could get out of control. But if you really think about it, it's much worse than that.

NPR: Much worse than *Terminator*?

MR. ROEDERSHEIMER: Much, much worse.

NPR: How could it possibly—that's a moonscape with people hiding under burnt-out buildings and being shot by lasers. I mean, what could be worse than that?

MR. ROEDERSHEIMER: All the people are dead.

NPR: In other words, forget the heroic human resistance. There'd be no time to organize one. Somebody presses ENTER, and we're done.

Yudkowsky believes superintelligence must be designed from the start with something approximating ethics. He envisions this as a system of checks and balances so that advanced AI growth is auditable and controllable, so that even as it continues to learn, advance, and reprogram itself, it will not evolve out of its own benign coding. Such preprogrammed measures will ensure that superintelligence will "behave as we intend even in the absence of immediate human supervision."[12] Eliezer calls this "friendly AI."

According to Yudkowsky, once AI gains the ability to broadly reprogram itself, it will be far too late to implement safeguards, so society needs to prepare now for the intelligence explosion. Yet this preparation is complicated by the sporadic and unpredictable nature of scientific advancement and the numerous secret efforts to create superintelligence around the world. No supranational organization can track all of the efforts, much less predict when or which one of them will succeed.

Eli and his supporters believe a "wait and see" approach (a form of satisficing) is a Kevorkian prescription. "[The birth of superintelligence] could be five years out; it could be forty years out; it could be sixty years out," Yudkowsky told us. "You don't know. I don't know. Nobody on the planet knows. And by the time you actually know, it's going to be [too late] to do anything about it."

YOU CAN'T JUST UNPLUG IT

Perhaps because of Hollywood or perhaps because humans just cannot help but anthropomorphize smart computers, many equate superintelligence with sentient robots. Whether friendly or murderous, they are an example of what is known as "embodied AI," superintelligence within a single robot or server. The ability to simply flip such a robot's on-off switch might be a comforting thought. However, superintelligence would likely be able to wreak havoc or even end humanity without ever leaving the cloud.

Another version of superintelligence, "disembodied AI" doesn't have a discrete, tangible location inside a single machine but exists as a program running on interconnected computers located anywhere. Yudkowsky fears that such superintelligence, even if initially created in a lab, could easily escape from its home server into the ever more connected world. Once in the wild, it would be inexorably intertwined within the global computer network, becoming nearly impossible to eradicate.

Eliezer argues that attempts to contain potentially dangerous artificial intelligence will fail. It will always find a way to escape, even through manipulation. To prove this, Eliezer devised an "AI box experiment," a simple game in which he played the role of AI and another person played the role of a human gatekeeper. Eliezer bet that in less than two hours, using only text messages, he could convince the gatekeeper to connect him to the Internet. Even with the gatekeeper solidly resolved beforehand (and with money at stake), the gatekeeper folded and the AI (Eliezer) won three times out of five.[13] This experiment simply pitted one human against another; in a real-life scenario, AI with superhuman intelligence would be far more persuasive and persistent in escaping its confines.

As the genesis of superintelligence approaches, Yudkowsky believes that fear, greed, biases, and national security priorities will overwhelm the caution he has preached. Eliezer told us that humanity's best hope is to perhaps create one highly funded, highly secure, multilateral effort to develop a friendly superintelligence with himself (or perhaps another futurist he approves of) at the helm. The work of this massive global Manhattan Project would be explicitly "for the benefit of humanity internationally." It simultaneously would ban, starve, or simply outpace other, less-well-thought-out efforts to develop superintelligence. Once created, this friendly AI would be unleashed to attack and destroy any competing efforts, ensuring that the only superintelligence in existence would help, not destroy, humankind.

Yudkowsky rejects the idea that a superintelligence should, or could, be tailored to parochial national security interests, believing instead that any solution must be considered at the human species level. "This stuff does not stop being lethal because it's in American hands, or Australian hands, or even in Finland's hands," he told us, mildly annoyed. "If China manages to build an AI, and that thing is actually safe . . . and they can safely give orders to their AI along the lines of 'melt the computers of all American AI projects' . . . China should do that. And we should be grateful. Because of the sheer deadliness of this stuff, if they can actually make it safe, if they can actually pull it off, then I am down on my knees, bowing to them in gratitude because at least we didn't end up as [molecular feedstock]. That's how serious this is."

Yudkowsky bemoans the fact that no governments have even hinted at an interest in such an effort, that others cannot see the urgency of his solution leaves him exasperated. He lets it rip. "The reason we're fucked is the result of an arms race for machines that we won't know how to control." He ventures to say that an alternate solution could include "someone like Bill Gates realizing that we're fucked . . . and why it's hard, and setting up their own smaller project." But he isn't optimistic. "Probably we're not going to get even that, so, yep, by default, on the current trajectory, we are fucked."

Yudkowsky takes a highly pragmatic approach to the world and communicates with others according to what are called Crocker's rules, one of which holds that the most efficient path of communication is always the best, even at the expense of social niceties. This blunt communication style—some might call it abrasive—has alienated potential allies over the years. While he is viewed by some as an oracle of reason, others call him a fearmonger with misplaced hyperbole and an inflated sense of self-importance. He wrote in 2007, "Lonely dissent doesn't feel like going to school dressed in black. It feels like going to school wearing a clown suit."[14]

Bullish tech experts point to how AI has already and will continue to benefit society. Ginni Rometty, the CEO of IBM, says, "In the future, every decision that humankind makes is going to be informed by a cognitive system like Watson, and our lives will be better for it." Another IBM executive discounts the idea that Watson could become a threat, because "the only data [Watson] has access to is the data we provide it with. It is not capable of going out on its own and creating—in some iRobot-type of form—its own data construct."[15] IBM also has good reason for touting the safety and promise of its technology: Watson is anticipated to generate $10 billion in revenue for IBM by 2023.

Noted futurist and AI cheerleader Ray Kurzweil welcomes the advance of superintelligence and believes man and machine will become one in a happy marriage he calls "the singularity." He envisions an increasingly intertwined symbiosis between technology and human thought and perception. But even Kurzweil agrees with Yudkowsky's goals for friendly AI and the perils of uncontrolled machine intelligence. "We have the opportunity in the decades ahead to make major strides in addressing the grand challenges of humanity. AI will be the pivotal technology in achieving this progress. We have a moral imperative to realize this promise while controlling the peril."[16]

In 2013, 550 of the world's leading AI experts were polled about the timing and risks of human-level artificial intelligence. Collectively, they predicted there is a 10 percent chance that human-level AI will be developed by 2022, a 50 percent chance it will exist by 2040, and a 90 percent likelihood that human-level AI will be with us before 2075. The "experts expect the systems to move on to Superintelligence less than 30 years thereafter."[17] Further, when asked about the long-term effects, 54 percent predicted that superintelligence would have a neutral or negative impact on humanity (19 percent anticipated an existential catastrophe, 15 percent said it would be bad, and 20 percent thought it neutral).[18]

Over the years, warnings like Eliezer's about the threat from artificial intelligence have begun to have an impact in the AI community. DARPA (the Defense Advanced Research Projects Agency), whose mission is to ensure that the U.S. military is "the initiator and not the victim of strategic technological surprises,"[19] launched a program for "explainable AI." "Machine learning and deep learning algorithms . . . we don't fully understand today how they work." The new explainable-AI initiative "will give the human operator more details about *how* the machine used deep learning to come up with the answer."[20]

In 2015, business tycoons Elon Musk and Sam Altman created the OpenAI Institute, a nonprofit company that focuses on researching AI. Musk and Altman believe that by making all of OpenAI's findings open-source and funding it by private donations, eliminating the need for financial return, they can ensure that AI will be developed for the benefit of all people, not for self-interested or destructive aims. They and others are so convinced of its importance that they have committed a total of $1 billion toward the initiative. Still, Musk says he is "increasingly inclined to think that there should be some regulatory oversight, maybe at the national and international level, just to make sure that we don't do something very foolish."[21]

WEAK AI: TRANSFORMATIONAL, FROM WALL STREET TO WEAPONS

Agree with Yudkowsky or not, artificial intelligence is already an inexorable part of society. Modern automobiles, smartphones, and even some appliances rely on AI. It runs aspects of the economic, health care, and judicial systems. *The Intercept*, the online news group powered by the Snowden leaks, has released documents they claim are from the NSA which show AI may even play a role in creating the lists of targets for secret drone operations. While Yudkowsky, Musk, Hawking, and others are focused on what they see as an existential risk from a rampant superintelligence in the future, others are concerned about the current status of weak AI and how it will continue to impact society.

The stability of the global economy relies on weak AI every day. Large-scale institutional investors have connected weak AI software directly to stock trading floors around the globe in pursuit of new profits. Using complex algorithms, computers now act as high-frequency traders, buying and selling shares on a massive scale in fractions of a second. These robo-traders analyze sales patterns, balance statements, regulatory filings, news reports, social media, and much more, and then buy and sell stock with little to no human intervention. In an increasing number of instances, humans are not even sure why the computer made the trades. While highly secretive, high-frequency trading is no sideshow oddity, it is consistently one of the most profitable trading strategies and accounts for 50 to 80 percent of daily global stock market volume, 1.6 billion shares bought per day in U.S. equities alone.[22]

But this effortless, automated trading can have a downside: weak AI can be fallible. On May 6, 2010, the U.S. stock market suffered a meltdown of epic proportions. During this "flash crash," $1 trillion of value was wiped from the stock markets in less than ten minutes. Then, about as quickly as they slipped, the markets recovered. Later investigations suggested that errors in the autonomous algorithms of high-frequency traders were at least partly to blame.

While AI has fundamentally shifted life on Wall Street, so too is it changing Main Street in new and potentially profound ways. Andrew Ng, the father of deep learning (a branch of machine learning that attempts to mirror human cognition), believes that focusing on the threat from superintelligence is misplaced. Ng is behind Google Brain, one of the most aspirational AI systems yet, and he feels that "worrying about AI evil Superintelligence today is like worrying about overpopulation on the planet Mars. We haven't even landed on the planet yet."[23] Ng believes that resources and time would be better spent on more pragmatic realities, such as the job displacement that he believes weak AI will cause.

A 2013 Oxford University study claimed that 47 percent of all U.S. jobs are at high risk of being automated within twenty years.[24] In addition to blue-collar jobs, the study suggests that an increasing number of knowledge workers will also be at risk. Ever-improving weak AI married with large bodies of data will obviate the need for stockbrokers, medical diagnosticians, information technology support staff, and travel agents. It suggests that even lawyers, doctors, and investment managers will soon find themselves competing and losing to weak-AI software that can more rapidly assess the relevant data and make decisions with "deep, specialized, and often tacit knowledge."[25] A 2013 McKinsey Global Institute study predicts that weak AI will depose 140 million full-time knowledge workers worldwide.[26]

The idea is certainly not new. In 1933, John Maynard Keynes predicted widespread unemployment "due to our discovery of means of economising the use of labour outrunning the pace at which we can find new uses for labour."[27] Technological evolution is part and parcel of the history of humanity; the introduction of the wheel, gunpowder, the steam engine, the car, the adding machine, have all led to systemic societal transformation. Most would contend that the world is better for it. The much-derided "buggy-whip maker," who lost his job with the advent of the horseless carriage is a popular symbol of the workers left behind by technological progress. Forgotten in that analogy are the buggy-making firms (and presumably their employees) who were able to pivot and profit in the new auto industry.[28]

Ng and others think weak AI will be far more disruptive. Losing 47 percent of jobs within twenty years is immensely more rapid than any technology-induced job loss the world has ever seen before. Artificial intelligence may well be a different kind of technological evolution, one in which workers made redundant by weak AI are unable to rejoin the labor force because there are simply no more jobs to fill. Weak AI could signal the end of the ability to earn, if not the end of labor. This inevitable loss of not only jobs but of purpose and a sense of productivity is perhaps more insidious than currently considered. What will be the role of humankind when machines can do the vast majority of jobs? What does a society look like when the labor force can no longer earn?[29]

In 1932, every fourth U.S. household had no breadwinner,[30] and the unemployment figures in Europe and Russia were just as glum. Franklin D. Roosevelt saw unemployment as the greatest threat to the nation since the Civil War. "There had never been a time when our institutions were in such jeopardy."[31] FDR was right. Unemployment is corrosive to government stability and calls for remarkably deft leadership, lest the nation collapse. In 1932, the U.S. responded with the New Deal. Western Europe responded with Fascism and the imminent rise of Nazism, Russia deepened into Stalinism and five-year plans.

Large-scale unemployment in the current era is no less disruptive and dangerous. The rise of radical Islam throughout the Middle East, the rise of narco-terror in Latin America, and spikes in inner-city gun violence in the United States all have strong correlations with the very low employment rates of young men in those areas.

AI also has revolutionary implications for warfare, and a deadly new arms race is already underway. Described as the third revolution in warfare, autonomous weapons can select targets and fire without any human intervention. South Korea employs the Super Aegis II, an autonomous robotic turret that can identify and fire on threats up to 3 kilometers (1.8 miles) away. Great Britain's "fire and forget" Brimstone missiles can hunt down targets in a predetermined area without any human direction. The U.S. Navy recently released a video demonstrating that autonomous unmanned "swarm boats" can overtake and destroy hostile ships without any human intervention. It also quietly canceled a contract to build a fleet of carriers deploying unmanned combat aircraft; the pilotless squadrons had been planned around the autonomous killer drone developed by Northrop Grumman, the X-47B.[32]

William Roper, the senior U.S. defense official charged with building the next-generation fighting force, puts AI at the core of his strategy. He hopes that AI will complement future warfighters, who would "quarterback" teams of autonomous war machines.[33] But Roper understands how hard that is going to be. "The thing that's scary . . . is that there's no reason that the processing time and reaction time from those [AIs] will not continually speed up beyond the human ability to interface with it." Other nations with autonomous weapons may not be inclined to keep humans in the "kill chain," the wartime decision-making hierarchy, once they become the slowest part of the process. That worries Roper. "There's going to be a whole level of conflict and warfare that takes place before people even understand what's happening."[34]

COMPUTING THE CASSANDRA COEFFICIENT

Daunting as they are, the challenges of weak AI pale in comparison to the extinction-level event that Yudkowsky warns could result from unrestrained superintelligence. While experts estimate that, at best, its invention is a decade or more away, that does not mean we have decades left to plan. Yudkowsky, Musk, Gates, Hawking, and others suggest that we need to begin preparing now. How concerned should we really be about this threat?

Yudkowsky's warning is certainly affected by Initial Occurrence Syndrome. Never before has our species encountered something of greater intelligence, so we are ill equipped to believe, let alone foresee and plan for, the complete scope of the problem. At the same time, Complexity Mismatch suggests that decision makers may not be able or willing to digest and distill the issue and its possible solutions. Moreover, the issue certainly seems outlandish to many, the stuff of sci-fi movies (but likely won't in even ten short years). Killer robots? Machines taking over the world? Does the seemingly fantastical nature of superintelligence result in a dangerous and dismissive bias against the issue? Certainly it has.

The audience for Eli's warning suffers from diffusion of responsibility. There is no person or office in the U.S. government or any international organization who is responsible for saving the world from superintelligence gone rogue. Yudkowsky's concern is so novel and perhaps so widely dismissed that there is not even an accepted scientific convention on AI safety protocols. A legitimate attempt at regulation would have to be done multilaterally, at the highest levels of every industrialized government (likely via UN treaty) and would require intrusive and likely surprise inspections of both government and commercial labs.

Yudkowsky is a data-driven expert who has devoted his life to preventing this disaster. He exhibits all of the Cassandra characteristics and is missing the Cassandra weakness of low social power. Even so, his singular focus and his adherence to alternative social norms has alienated others who might serve as advocates, even those who might play a role in a solution. Is our Cassandra's off-putting personality an insurmountable obstacle?

Finally, many of Eliezer Yudkowsky's critics have vested interests against regulations or other efforts to slow the development of AI. From those in industry to those in government, superintelligence holds the promise of an unmatched competitive edge. Can they continue to be objective parties to the discussion, or have they grown biased against the threat?

Yudkowsky's suggested solution, a global Manhattan Project to develop safe AI, would be one of the most incredibly complicated multilateral bureaucratic solutions we could imagine. An alliance of the United States and select allies may be a more reasonable solution, but nothing of this sort has even been considered. We have been warned that we have one chance, and one chance only, to get it right. Perhaps we should pay attention to Eliezer Yudkowsky before we open a door we can never again close.

Yudkowsky and Artificial Intelligence

CASSANDRA CHARACTERISTICS	THE WARNING	THE DECISION MAKERS	THE CASSANDRA	THE CRITICS
High	X	X		X
Moderate			X	
Low				

CHAPTER 11

The Journalist:
Pandemic Disease

The more we win, the more we drive infections to the margin of human experience, the more we clear the path for possible catastrophic infection.... We are caught in the food chain, whether we like it or not, eating and being eaten.

—WILLIAM MCNEILL, PROFESSOR EMERITUS OF HISTORY, UNIVERSITY OF CHICAGO

The virus was so deadly that the entire populations of some Alaskan Inuit villages were killed before word made it out that they needed help. Sled dogs, now abandoned, broke their tethers, roamed wild, and survived on the frozen human remains scattered where the sick dropped in the streets and the houses of the town.[1]

The bug emerged after jumping from birds to pigs, then infecting French soldiers on the battlefields of World War I. It took twenty-four months to make it to Alaska, and before it was over, the Spanish flu (1918–1920) had killed more people than the Great War.

In 2009, an outbreak of a nearly identical virus spread rapidly through Mexico and the southwestern United States. The World

Health Organization (WHO) declared a pandemic; President Obama declared a national emergency. Four strains of influenza had incubated and reassorted in their animal hosts: American ducks and chickens, Mexican and Asian pigs, even northern sea otters of Washington state, before jumping to humans in Veracruz, Mexico. Within twelve months, the virus had struck every nation of the world.

Spanish flu was not on R.P.'s mind in the mid-1990s. Overseas, after meetings with foreign defense officials, he escaped to a bar to test the local beer. An older man came in behind him, sat down, and sparked up a conversation. Beyond his short haircut, nothing was notable other than the man's steady persistence at conversation. They shared a beer's worth of small talk, then he put his hand on R.P.'s elbow, lowered his voice, and confided that he had followed him into the bar.

The man hoped to trade highly classified secrets for U.S. asylum. An active-duty, senior military scientist working on the covert weapons program of a state we had long suspected of violating the Biological Weapons Convention, he wanted out. This scientist was terrified of the impact from the germs he had been ordered to design. Ebola . . . anthrax . . . chimeras . . . recombinants. The beer got suddenly less interesting.

With some calls back to D.C., R.P. arranged for the appropriate agencies to vet and handle this walk-in. Having worked the bioweapons and disease portfolio on the National Security Council, he was amazed by the disclosure and was curious to hear what our biological weapons experts thought. Our scientists at the U.S. Department of Defense were startled and found the information hard to believe. The source told a tale of hidden labs mixing and matching bacterial and viral DNA to brew up the diseases of doctors' nightmares: a superbug that's transmissible by air (perhaps via a sneeze) and kills a large percentage of the infected. A new disease that spreads like the flu and kills like smallpox could wipe out a billion people in six months with no reason to think it would stop there.

Our brightest minds felt the scientific tools at that time were simply not available to create these chimeras. They didn't think the walkin's nightmare Andromeda strains could be real. They weren't sure if *he* believed what he was saying, but *they* didn't.

That was then.

MELTING FROM WITHIN

R.P.'s run-in occurred during an era when pandemic diseases suddenly regained prominence in the White House and the national security agencies. In 1995, in what used to be called Zaire, a terrifying virus reemerged after slumbering since 1976. Ebola crept out from deep within African forests and killed 80 percent of the unlucky infected. They died horrifically, in puddles of blood, as if they had melted from within. And for more than a decade, the world had been gripped with fear of a new infection nobody had ever seen. HIV/AIDS was decimating gay communities in the United States and propagating rapidly all over the globe.

As happens more often than those in power want to admit, government concern was partly catalyzed by Hollywood. In March 1995, the same month as the Ebola emergence in Zaire, Dustin Hoffman, Rene Russo, and Morgan Freeman starred in the blockbuster *Outbreak*. An African monkey, carrying a deadly, hemorrhagic virus, is smuggled into the United States. The fictional virus crosses from its animal host to infect residents of a California town, mutating and spreading person-to-person like the flu. Lacking a coherent pandemic response plan, the U.S. government orders the town exterminated to contain the virus and cover up the outbreak. Thankfully, a scrappy band of scientist-heroes finally create a cure and save the town from military annihilation just in the nick of time.

Around the same time, President Clinton read a newspaper article about Zaire's horrific Ebola outbreak. In a handwritten scribble in the margin of the article he asked Tony Lake, his national security advisor, what the federal plans were in the event of a real outbreak: "Tony/NSC: Are we ready for something like this?" The President's question set off a scramble in the White House to answer more immediate questions: Do we even consider this a threat? Who in the NSC works on deadly disease outbreaks?

The scribbled question landed in Dick's office. After deciphering the notorious scrawl, we realized the President had asked a question both prescient and horrible, because we knew we were prepared to do dismally little. The potential Cassandras we will soon meet say we are not much better prepared today.

The Grim Reaper's favorite disguise is disease. Disease makes other disasters look trivial. More human lives have ended by bacteria and viruses than every other kind of catastrophe combined, its constant presence masking its destruction. Even war flags behind. The Civil War killed 750,000 Americans, more than all other U.S. wars combined, and the top killer of American soldiers in the Civil War was disease, not combat. In 2014 alone, AIDS killed 1.2 million people; it has killed 39 million people since the pandemic began, a little more than the entire population of California. Another 36 million people alive today have HIV/AIDS, including 2.6 million children.

Disease is now and has forever been humankind's greatest killer, and it certainly is not being ignored; health care is one of the world's biggest industries. So how could disease get any worse? Is there really a Cassandra here?

KILLING HALF THE WORLD

Beginning in 1918 and lasting less than three years, the Spanish flu epidemic killed up to 5 percent of the earth's population. Death often came quickly; the infected often felt fine in the morning but were dead before the next day's light. The disease was so deadly that it burned itself out: it killed victims so fast that they didn't have time to infect many others. Around 30 percent of the world population caught it, and fifty million people died, almost seven hundred thousand in the U.S. alone. By comparison, World War I killed fewer than nine million people; World War II, fifty-five million.

Almost a hundred years later, after numerous studies, including sequencing the virus's RNA, scientists still do not know exactly why the Spanish flu was so deadly, and that ignorance obviously makes experts uncomfortable. What the science does say is not reassuring. It says that this deadly flu is nearly identical to H1N1, commonly referred to as swine flu. H1N1 is a subset of the influenza A virus, one of the two categories of seasonal flu virus millions of Americans get every year. In fact, Spanish flu infected so many in 1918 that it is the genetic Adam and Eve of nearly any modern pandemic flu strain. These variants are so genetically similar that the devastating virulence of Spanish flu could return from a change in as few as three different proteins encoded by its RNA.

New mutations of influenza have yet to match the lethality of the 1918 pandemic, but on several different occasions the virus got terrifyingly close. In 1957, a flu virus that had not been seen for sixty-five years was discovered in Asia. Because of its long dormancy, most of the world had no immunity. In two waves of deaths, nearly 1.5 million people were killed, including seventy thousand Americans. The 2009 influenza pandemic killed almost three hundred thousand people. The virus was a reassortment of four different known strains of influenza A, two from pigs, one endemic in birds, and one endemic in humans.

Influenza is amazingly adaptable. It changes lethality and transmissibility quickly and jumps from animal to human more readily than any other disease. New flu mutations emerge *daily*, some proving more contagious than others. Tests of patients exposed to the deadly 2009 flu outbreak show that the virus evolved each day as the patient was exposed to antiviral drugs.

Jeffery Taubenberger, with his colleague Ann Reid, was first to sequence the Spanish flu. He heads the effort at the National Institutes of Health to combat our next flu pandemic. Taubenberger says, "If the records are any indication, yes there is another pandemic coming—the question is when."[2]

Flu is by nature highly contagious, but it also circulates best when it is not highly lethal. Killing the host too quickly prevents maximal spread of the virus. Taubenberger fears a flu endemic to one animal could make the leap, the "epizootic shift," and become transmissible among humans. Such a virus could be less lethal in its host animals (who survive to keep spreading the virus), but be highly lethal to humans. A version of H5N1, the bird flu, can infect humans and kills with breathtaking lethality, over 60 percent dead. The good news is that H5N1 is not yet transmissible from human to human. Could this deadly flu gain the ability to spread itself from person to person?

In 2011, Ron Fouchier, from the Erasmus Medical Center in downtown Rotterdam, crafted a series of experiments to mutate highly lethal H5N1 into a form contagious by air. Just five single mutations allowed H5N1 to bind with cells in the human respiratory tract (thereby making it contagious by air, sneezes, and dirty doorknobs, etc.). Using ferrets as incubators, and their noses as makeshift Petri dishes, Fouchier rapidly moved infected sputum from ferret to ferret. In a period of weeks, he created a bug as transmissible as the Spanish flu but potentially up to twenty times more lethal. Such a bug could lead to an extinction-level event for the human race.

Fouchier was roundly criticized for launching such a dangerous study in a working hospital, in a crowded city, with arguably less than perfect protections. He did it without complicated tools, available in nearly any laboratory and to consumers. And he decided to publish his results to give the world a step-by-step manual, steps that could be taken in nearly any lab to make his superbug.

The U.S. National Science Advisory Board for Biosecurity tried to block Fouchier from publishing key aspects of his research.[3] The Dutch government ruled that Fouchier could not publish this information because that amounted to the export of deadly weapons, requiring a special export license. Fouchier agrees that the results of his experiment are "very bad news," but he hopes that, by making his research public, scientists can study and remain one step ahead of mutations that could create pandemic airborne viruses. Publish and perish?

DID THE VIROLOGIST CRY WOLF?

Robert G. Webster is a virus hunter. He has mapped how the influenza viruses that cause human pandemics brew in chickens, pigs, ducks, and other animals. His discovery upended the previous theory that new pandemic flu strains emerged from genetic reassortments of human flu viruses. Humans, then, might have little to no resistance to these animal-derived strains, any individual's immune system naked and vulnerable.

Webster's discovery certainly scared him. He has been sounding the alarm about pandemic flu for decades. There are "about even odds at this time for the virus to learn how to transmit human-to-human," Webster says.[4] "We cannot afford simply to hope that human-to-human spread of H5N1 will not happen and that, if it does, the pathogenicity of the virus will attenuate. SARS was interrupted by early case detection and isolation, but influenza is transmissible early in the course of the disease and cannot be controlled by similar means."[5]

If Fouchier's pandemically engineered H5N1 had breached his laboratory, or if something similar emerges naturally, the resulting damage to civilization is hard to fathom. Even if H5N1 follows the Spanish Flu percentages—30 percent infected and a lethality of 2 percent—Fouchier's superbug would kill 42 million people. However, Webster thinks that is a foolishly low estimate. Why would the lethality decrease when Fouchier just proved the virus can become airborne without weakening? Webster thinks 3.5 billion could die. "Society just can't accept the idea that 50 percent of the population could die. And I think we have to face that possibility. I'm sorry if I'm making people a little frightened, but I feel it's my role."[6]

Dr. Webster told us that scientists can predict that a pandemic flu will likely strike at some point, but the timing and type are nearly impossible to foresee. "The difficulty occurs because we can't make accurate predictions. We see the potential of this approaching, and we prepared for H5N1 . . . then nature gave us H1N1. The public started thinking, 'These guys really don't know what the hell they're doing.' Then H7N9 came along and moved faster in humankind. We can't predict when one of these things will learn to jump from human to human."[7]

Like other Cassandras, Dr. Webster looks to data to substantiate his warnings and does not understand why his conclusions aren't obvious to everyone. "The information has been building up for years and years. The scientific community has gotten the message that the influenza virus has the capacity to spread from wild birds into humans and cause devastating effects," he told us. "But the public starts doubting us because we can't predict when it will happen. That is the frustration."

He believes decision makers have a hard time accepting new costly solutions after epidemiologists' expensive warnings that H5N1 was going to strike in 2009, when, in fact, it was an H1N1 epidemic that struck and killed three hundred thousand worldwide. "In '09, we spent vast amounts of money stockpiling oseltamivir IX [Tamiflu]. They predicted H5N1 and spent huge amounts with industry for an insurance policy we did not end up having to need." In the years since, decision makers believed the Tamiflu stockpiling was a wasted effort. As Dr. Webster said, this has tarnished the reputation of the field of public health, discrediting experts who suggest that government needs to better prepare for pandemic disease. Investigative commissions even investigated the WHO and other health services over allegations of collusion with Tamiflu's manufacturer, Roche.

While more recent studies have shown that Tamiflu actually saved thousands and reduced the death count by 25 percent, the reputational damage has been done. Budget allocators appear less willing to fund pandemic preparedness. "It is really showing now as we learn more about the Zika virus." Dr. Webster told us, referring to the mosquito-borne virus that has ripped through Latin America in 2016. "This is a most terrifying virus, and the [public health] funding is just not there."

TWO-TIME CASSANDRA? LAURIE GARRETT

Laurie Garrett should have no trouble drawing an audience.

She has thirty years of data and experience to describe how the next lethal and highly transmissible disease will very likely kill hundreds of millions of people. She has studied and advised every organization charged with protecting us from the next global plague and is quite sure they will still be caught largely flat-footed. She tells a frightening story with extraordinary confidence: the powers-that-be "just don't seem to get it." Surrounding her is an air of profound frustration; she knows society is unprepared for the next superbug. For her writing, broadcast work, and advocacy, Garrett is the only journalist to have won all three of the coveted "Big P's," a Pulitzer, Polk, and Peabody prize. Yet few seem willing to listen to, let alone act on, Laurie's warnings.

Not simply a journalist who reports on public health, Garrett has a distinguished background in science. She began her Ph.D. at Berkeley in the Department of Bacteriology and Immunology and did laboratory research at Stanford before discovering a love for journalism. Happy circumstances introduced her to a radio show where she was featured as a guest. "It became a regular gig," she told us, and the opportunity eventually led to her own show called *Science Story*. Her first effort as a journalist was quite successful; she won the Peabody Award for the best radio broadcast of the year. "That was a lot more fun and easier than killing lab mice, and I wondered if this is what I should be doing."[8] And off she went.

Covering a war was the hardest thing a journalist could do, Laurie was told, so she moved to Africa to become a war correspondent in the late 1970s. "I had seen poverty in the United States, but that was nothing like the poverty in Africa," she recalled to us. "I had never seen massive child death; I had never seen children die in my arms. And the number-one killer wasn't war; it was measles."

Laurie was staring into the public health vacuum in the developing world, and she felt a responsibility. "It never really occurred to me what global inequity really was. Not until you see a totally preventable disease cause hundreds of children to die in front of you, you don't know what inequity means. . . . I was so simultaneously saddened and indignant that I knew this would be the major frame of my life going forward." Yet while millions in the developing world were dying of infectious diseases, universities in the United States were shutting bacteriology departments on the assumption that infectious diseases were conquered.

Returning to the U.S., Garrett became a science reporter for NPR out of San Francisco and ran into the middle of another health crisis. "Our office was in what was called the Meat Market because during the day it was a warehouse district, and at night it was a massive pick up scene for gay men. . . . There was this weird cancer that was killing young healthy guys, turning them into sticks." The first official designation was "Gay Associated Cancer. . . . After three months [of watching these gay men get sick and die] I was convinced it was an infectious disease, but there were all sorts of other theories." Laurie was terrified by what she saw and felt compelled to raise the alarm. "I had no other choice. I decided that I needed to report this story." She became the first national journalist to cover and warn about what we now know as HIV/AIDS.

It was a time of national panic. "Ryan White's school got burned down,"[9] and "national-level politicians called for putting AIDS patients into camps and tattooing infected men," Garrett remembers.[10] "I got a lot of complaints from my higher-ups and the FCC about the explicit biology of my stories. It became very uncomfortable at NPR because [an NPR leader] said things like, 'if I hear one more thing about dying faggots, I'm going to puke.' I was fighting a battle on all fronts, and did not feel supported by my institution. But I was so inspired by the heroes I met out there . . . not just the doctors and nurses, the social workers doing needle exchanges and talking to prostitutes and gay hustlers, and who didn't know if their own life was at risk. It was so remarkable."

Laurie was in the vanguard of warning and teaching about HIV/ AIDS and responding to it. As HIV became a global crisis, Garrett became a global force. She started reporting on the policy conversations regarding the response to HIV and other public health issues. "I was beginning to see how policy moved and change could happen." Garrett then decided she wanted to help drive policy.

In the late 1990s, Ambassador Richard Holbrooke was President Clinton's Permanent Representative to the United Nations. R.P. was leading his efforts to bring HIV/AIDS into the national security dialog of the United States and the international community to elevate the issue's importance and break through bureaucratic barriers. Though we met Garrett as a journalist, it soon became abundantly clear that she was as much of an expert on the science, sociology, economics, and bureaucracies of disease and disease prevention as anyone we would ever meet.

Laurie was a powerful and mobilizing influence on the subject of HIV/AIDS, but years would be needed to overcome the fear, misinformation, and bias that hindered our efforts to effectively deal with the epidemic. Sadly, so many lives were lost due to ineptitude and inaction that the initial decades of global response can be labeled as nothing more than a failure. Laurie's warnings about HIV/AIDS were prescient, but it took far too long for decision makers to grasp the gravity of the problem and make a concerted effort to implement a response.

After her work with HIV/AIDS, Laurie spent decades tracking disease outbreaks, her passion continuing to drive her career. She tells us that she has seen a massive uptick in the emergence of new diseases and the reemergence of old ones. She believes that this is not a coincidence, that the world is out of balance and deadly microbes are on the rise.

The introduction of penicillin in the 1940s and the start of the antibiotic age is widely considered one of the great advances in public health. Previously, a diagnosis of meningitis, pneumonia, tuberculosis, or even an infected cut or scratch was akin to a death sentence. Hospitals were full of people simply waiting to die, doctors unable to do anything to save them.[11] Penicillin ushered in a wave of optimism: the development of increasingly powerful antibiotics, the near-eradication of many diseases, the hope that man had achieved complete mastery over bacteria.

In the past twenty years, that feeling of mastery has turned to growing fear, fear that antibiotics are losing their efficacy. The excessive and improper use of antibiotics has imparted resistance on the bugs they used to kill. Higher doses and more powerful antibiotics have ushered in more powerful, multi-drug-resistant pathogens of all types, bacteria, viruses, fungi, and protozoa. Already, millions have died from drug-resistant microbes.[12] The WHO warns, "This serious threat is no longer a prediction for the future; it is happening right now in every region of the world and has the potential to affect anyone, of any age, in any country. Antibiotic resistance—when bacteria change so antibiotics no longer work in people who need them to treat infections—is now a major threat to public health."[13]

While antibiotic resistance has historically victimized poor countries in particular, due to their weaker public health infrastructure, the threat to the United States is now stark. In May 2016, the U.S. public health service announced that the first bacteria resistant to all antibiotics, even the last lines of defense, had been found in the country. First isolated in China the year before, individuals unlucky enough to be infected by this strain of bacteria can hope for little more than the patients of the early 1900s: that doctors can make them as comfortable as possible while the disease ravages their bodies.

INVISIBLE ANTHRAX

Garrett is not concerned just about the diseases that emerge and re-emerge from nature; she is equally concerned about what man can do. After the September 11th terrorist attacks, she realized that if terrorists were willing to commit suicide, it stands to reason that they would not mind being infected by the bioweapons they were releasing. That solves what had long been considered the critical challenge for terrorist groups interested in using these agents: dispersal. Effectively dispersing bioweapons (without infecting the persons dispersing them) had long been considered very difficult, likely beyond the means of a terrorist group. But that problem goes away if a terrorist doesn't mind dying with the victims.

Terrorists' use of biological weapons is not just a fringe fear; it has already happened. The Japanese Aum Shinrikyo doomsday cult is a case in point. No intelligence agency had ever heard of Aum Shinrikyo when it released the chemical weapon sarin into the Tokyo subway in 1995, killing thirteen and injuring fifty-four more. Their founder, Shoko Asahara, sported a large Cheshire cat grin and a furry beard reminiscent of Jerry Garcia's; he had previously worked in a yoga studio as a massage and acupuncture therapist. He also built the largest nonstate biological and chemical weapons program ever seen.

After the same 1995 Ebola outbreak that seized President Clinton's attention, forty of the group's members traveled to Zaire disguised as aid workers to secure infected blood for biological weapons. Thankfully, they were unable to obtain any samples. Perhaps more frightening was their attempt to genetically engineer a more lethal form of anthrax.

Anthrax kills quickly, but anthrax is hard to kill; 85 percent of those unlucky enough to inhale anthrax spores die. It is an extraordinarily hardy species that can survive for centuries in the corpses of its victims. During World War II, the British exploded anthrax bombs over sheep on a small Scottish island called Gruinard. Forty years later the island was still contaminated. The British flooded it with three hundred tons of formaldehyde solution, and it still took four years until the island was considered clean enough for a human visit.

In 1989, scientists involved with the Russian biological weapons program published a research paper describing how to blend different strains of anthrax into a superstrain more resistant to vaccines and treatments. Aum sympathizers worked in a Japanese agricultural university where the Russian paper was reviewed. In 1992, the cult finally obtained a strain of anthrax, built a shoddy laboratory, and began its work. Aum eventually created copious amounts of a slurry infused with anthrax, and in 1993, to test its lethality, sprayed it from a rooftop in a crowded "village" (administrative district) in Tokyo on two separate occasions. Fortunately for the innocent people of this village, Aum had mistakenly obtained a veterinary vaccination strain of anthrax that posed no harm to humans.[14] Thankfully, the cult was better at massage and acupuncture than weaponizing anthrax. Yet the lesson of Aum should not be that it failed, but that its members were captivated by the ferocious power of certain superpathogens and attempted to use them to induce a pandemic. They are unlikely to be the last.

In chapter 16 we will assess the risks from the gene-editing tool called CRISPR. CRISPR's greatest threat may be its use as a tool to create lethal diseases. Even hardened experts like Dr. Webster, who fight every day against the power of nature as it seeks to create killer influenza recombinations, are flummoxed by the reassortments humans may seek through CRISPR. "How do we deal with this? I'm not sure I have the answers. You can't put the genie back in the bottle."

Dr. Webster told us, "We are going to have to work very hard to put in place the guidelines and safety measures necessary to protect society from those who want to do evil things, [because] it will be so much easier for them to do so."

Laurie Garrett feels similarly about CRISPR. Garrett has sat on the ground in Africa holding babies as they die from cholera; she has watched behind a paper medical mask as blood poured out of the eyes of Ebola patients. When we asked her about CRISPR, she paused and said, "Oh . . . my god! . . . There are serious questions if science should stop using it, maybe earlier we could have, but now it is too late. But now, my god . . . when you have competing weapons programs out there and kids are creating previously nonexistent microorganisms . . ." Accepting that the CRISPR technology will be around for the indefinite future, Garrett still maintains that a better public health surveillance-and-response system is the only way to protect against a man-made virus.

WHO'S WATCHING FOR THE COMING PLAGUE?

Laurie warns that while the risk of deadly microbes is increasing, governments' ability to detect and respond is weakening. In her best-selling books, *The Coming Plague* and *Betrayal of Trust*, she meticulously discusses the decline and often wholesale absence of a competent public health infrastructure. The inability to deal with these growing threats unveils the specter of catastrophe—and more so when governments are simply unwilling to take the threat seriously. It is the "same pattern everywhere of government mistakes," she told us, "of bigotry, of ignorance. It could be HIV in Africa or plague in India. Why are the same mistakes repeated over and over?"

Because the nature of the threat is so broad, Laurie's message is specific only in the solutions. She suggests that a wide range of events could plausibly cause a health catastrophe. However, she fears this sort of "broad risk and narrow response" message is ineffective at convincing decision makers and the public. "The most common question I get asked is: 'So what *is* the "coming plague"?'" Laurie thinks about public health threats constantly, but is resigned to the fact that nobody can predict precisely what it will be; a deeply unsatisfying answer for most audiences. "The threat could come from BL3 and 4 labs [government laboratories designed to house deadly infectious agents] in crazy places like Pakistan that are filled with the most dangerous pathogens on earth . . . or Zika, yellow fever, pandemic influenza . . . just name a microbe."

Like the Cassandras in the first half of this book, Garrett is regularly criticized for her views. "Every single time I try to draw attention to an outbreak, I can tell you the particular time when I will be attacked. It is always a white male who combines his attack with a comment about my appearance and usually something related to me being a female. When I was on NPR the other day, someone tweeted about how fat I am. That is how they discredit my point. You have to be wearing a skin of steel and be willing to take the barbs and arrows."

She also feels exasperated when she's criticized for not being more specific about a yet unknown pandemic. "If you draw attention to a problem you're accused of being a Chicken Little, but if you don't draw *enough* attention, you're blamed for ignoring the problem." She laments that in public health, "you never get credit for correctly predicting an outbreak" because implementing effective countermeasures blunts the impact of the disease, leading critics to believe "that you exaggerated the threat." Even a sophisticated crowd sometimes criticizes Laurie's warnings. "At the Council on Foreign Relations, I was told, 'You led us to believe H5N1 would be a devastating pandemic for humanity, but it never materialized.'"[15] (Recall that stockpiled Tamiflu was estimated to have blunted the mortality rate by about 25 percent.)

Warning of disaster is a complicated and tricky business, particularly when the solutions are expensive. And Laurie's solutions are expensive. "There is never going to be a way to have a microbe-specific defense infrastructure. You will never know in advance of every pathogen you will face. It will be one disease after the other. Fixating on one thing—bioshield, targeted intervention, vaccines microbe by microbe—just won't do it, especially when you have to deal with a microbe you didn't know existed."

Regardless of what deadly microbe comes up next—pandemic influenza, a man-made chimera, or multi-drug-resistant bacteria—Garrett insists that the world needs a greatly improved public health infrastructure to respond appropriately. "It always comes back to 'You need a public health infrastructure,' and no one wants to hear it. . . . It's not sexy. It doesn't inspire a lot of money or get a lot of kids mobilized. This is the common thread and solution-set, but it doesn't ring true or bring broad smiles to anyone outside of government and public health."

CASSANDRA'S EPIDEMIOLOGY

Scientific Reticence has led many to question these claims of a coming pandemic, but Webster's and Garrett's warnings are a reminder that "not sure exactly when it will happen" does not equal "it won't happen." What information would lead us to act? If we wait for only perfect and precise information, we court disaster.

Magnitude Overload and the Invisible Obvious may also play a role. As we noted in the beginning of this chapter, disease is and has been the primary killer of humans since the dawn of the species. Perhaps it is so ubiquitous that we have become blind to the Invisible Obvious that public health solutions, complex and expensive as they may be, can ward off what have historically been inevitable pandemics. Yet

Magnitude Overload impairs our capacity to grasp the true scope of the problem: millions, hundreds of millions, even billions struck down.

As disease respects no borders, Webster's and Garrett's appeal is directed at decision makers at every level of government, from the local mosquito-abatement official to the federal Public Health Service, the health services of other nations, and the important multilateral agencies like the WHO, UNICEF, and others. By the nature of the threat, there has to be a shared response throughout these strata of government, and that requirement most certainly adds to a diffusion-of-responsibility challenge. As strong as the debate has been about the will, ability, necessity, and value of the United States acting as the sole superpower, in the face of a pandemic, all questions would cease, and all eyes would be fixed on the Oval Office awaiting a solution.

Garrett believes that many entities necessary for a proper response are underfunded, poorly led, or simply incompetent. She makes the point that many leaders consider issues of public health only after they're faced with a crisis. Up until that moment of panic, Garrett believes most leaders pay lip service and offer minimal resources to public health challenges, the Band-Aid that we refer to in chapter 9 as satisficing.

Garrett herself is unique in our study in that she has already proven herself a Cassandra about the global explosion of HIV/AIDS, and she is now warning us about a related but different threat. In both instances, Garrett shows every one of the Cassandra strengths. For both Webster and Garrett, the expertise, reliance on data, questioning/skepticism, and sense of personal responsibility seem obvious in what we have learned above. Testimony to Webster's ability to be a first-principles and orthogonal thinker comes from his history of notable scientific discovery, and Garrett has shown it repeatedly, perhaps most clearly when she recognized HIV was something more than a cancer cluster.

One criticism leveled at both is that they have warned of disasters that haven't materialized. They are accused of being what Wall Street calls "permabears," pessimists who see a bogeyman, a pandemic, or a collapse around every corner. We will leave it to the reader to decide if their inability to be more specific in some of their disease predictions indicates false alarms or legitimate examples of accurate signaling.

We are not aware of any respected member of the scientific community who disagrees with Garrett's or Webster's work. As we discussed before, the failure to respond seems to be related not to doubting these potential Cassandras' warnings but rather to being unwilling to spend the heavy resources Garrett and Webster insist are necessary to build a global public health system to fight a foe we can't specifically pinpoint.

Perhaps because Webster and Garrett fit so well into the metrics of experts we ought to listen to, they are not ignored completely like some of the Cassandras we met in the first half of this book. Nonetheless, they and other experts in their field believe there is a dangerous lack of commitment and resources to the solutions Garrett and Webster call for. When the next pandemic strikes, all that will matter is the capacity of our public health system to detect and respond.

Garrett, Webster, and a Pandemic

CASSANDRA CHARACTERISTICS	THE WARNING	THE DECISION MAKERS	THE CASSANDRA	THE CRITICS
High	X		X	X
Moderate		X		
Low				

CHAPTER 12

The Climate Scientist: Sea-Level Rise

The waters rose and covered the mountains to a depth of more than fifteen cubits. Every living thing that moved on land perished—birds, livestock, wild animals, all the creatures that swarm over the earth, and all mankind. Everything on dry land that had the breath of life in its nostrils died.

—GENESIS 7:20–22

Symposium is a small Greek restaurant just off the Columbia University campus in the Morningside Heights neighborhood of New York City. It has all the makings of a New York ethnic restaurant, with exotic smells and a Greek staff churning out hummus and gyros, eggplant and baklava. Dr. James Hansen, wearing a denim button-up shirt and a leather-brimmed cap, arrived promptly at noon looking more like Indiana Jones than the planet's leading mind on climate change.

When we asked people for names of potential future Cassandras now among us, one name kept coming up: James Hansen. The problem was that he had already proven himself as a Cassandra, maybe *the* Cassandra of the last forty years. He put the issue of climate change on the world's agenda, over the yelling-and-screaming objections of two Bush Administrations and those in the oil and gas industries. He had risked his career but had ultimately been proven right. Now he was at it again. This time he is predicting rapid and massive sea-level rise, and again he is being criticized. If he is right once more, we are in very deep trouble, for sea-level rise will be faster and higher than even leading climate change experts believed possible.

Hansen is currently the director of the Climate Science, Awareness, and Solutions program at the Earth Institute of Columbia University.[1] Before his time with the Earth Institute, he was the director of the NASA Goddard Institute for Space Studies (GISS) from 1981 to 2013. There he built the center into a global leader in atmospheric modeling and climate change studies.[2] He was trained in physics and astronomy at the space science program of Dr. James Van Allen at the University of Iowa, but has focused his research on Earth's climate, an interest slowly evolving out of his planetary study of Venus's atmosphere. He was elected to the National Academy of Sciences in 1995 and was named one of the 100 most influential people on Earth by *Time* magazine in 2006.

We thought that spending some time with Hansen would give us an opportunity not only to probe him on his predictions about sea-level rise, but also to understand how being the Cassandra on climate change in the 1980s and '90s had affected him. We wanted to know how this Cassandra stood up to criticism, what drove him, how he approaches the issues he deals with, and more generally, who he is as a person. We quickly found out.

There was a time when the concepts of global warming and climate change were widely perceived as radical theories and fringe science. However, the year 1981 marked the beginning of a series of accurate predictions that remain "essentially unbroken" today.[3] Hansen published his first major paper linking global warming in the twentieth century, a global surface temperature increase of 0.4 degrees Celsius, with increasing CO_2 emissions and the greenhouse effect.[4] He wrote that the increase in temperature "is consistent with the calculated greenhouse effect due to measured increases in atmospheric carbon dioxide." He and his team at GISS stated with precise accuracy that the "potential effects on climate in the 21st century include the creation of drought-prone regions in North America and central Asia as part of a shifting of climatic zones, erosion of the West Antarctic ice sheet with a consequent worldwide rise in sea level."[5] The following year, he published another bombshell article that associated sea-level trends with global surface temperature.

These and subsequent studies helped Hansen become well-known within the scientific community for asserting the reality of human-caused climate change, but he failed to gain much traction with the wider public for most of the decade. He claimed that he has always been "shy, a poor communicator, and lacking in tact"[6] and that his warnings simply got obscured by the technical jargon. Then came the summer heat wave of 1988. Crops shriveled and died, livestock perished, cities throughout the country enacted water-consumption restrictions. Congress called hearings. Scheduled to testify before the Senate in June, he was determined to make a "pretty strong statement" using results from his new model, which gave greater precision in mapping global warming over time. This time he would abandon the jargon and go for "blunt."

In the run-up to the hearing, one of Hansen's colleagues at NASA Headquarters remarked that no respectable scientist would attribute the decade's warming to the greenhouse effect. Hansen remarked, "I don't know if he's respectable or not, but I know someone who is just about to make that statement." His Senate testimony was brief and to the point, lasting just five minutes. He stated that he was "ninety-nine percent confident" that Earth was getting warmer and that the warming was large enough to "ascribe with a high degree of confidence to the greenhouse effect." His computer model showed that the man-made greenhouse effect was already strong enough to increase the frequency of extreme weather events. After the hearing, surrounded by reporters, Hansen said, "It is time to stop waffling so much and say that the evidence is pretty strong that the greenhouse effect is here." The next day his words were the *New York Times* "Quotation of the Day."[7]

We were curious about what it took for Hansen to see global warming first, and in particular, if it required a lot of individual resolve or belief in himself. He referenced the physicist and Nobel Laureate Richard Feynman: "Well, I'm not a person who has a great deal of self-confidence. And I've considered that an advantage; Feynman says that 'You have to be very skeptical of any conclusion, especially your own because you are the easiest person to fool.' Often you start out, you want some answer, you think you know some answer and you try to persuade yourself. You have to question your conclusion and be very skeptical. That's the principal characteristic of a good scientist, to have the proper degree of skepticism. Even of yourself." He paused and then admitted his favorite quote is by Oscar Wilde who said, "When people agree with me, I always feel I must be wrong."

But people did not agree with him. He recalled meetings organized to dismiss what he had said to Congress. "There was a meeting in D.C., which was described as a 'Get Hansen' meeting: It was government and academic scientists wanting to conclude that I had gone way too far in my testimony. . . . Then the same thing happened [again]. Dr. Michael Schlesinger [from Oregon State University's Atmospheric Sciences and Climatic Research Institute] organized a meeting in Amherst, Massachusetts, several months later." Hansen then laughed, "It started on the day I had gone back to Washington to testify again, to the chagrin of the people in Amherst!"[8]

Hansen consequently missed part of the Amherst meeting, though he had been invited to attend, and the participants resorted to pot shots in his absence.[9] His colleagues argued that he didn't have enough information to back up his claim. They concluded, "It is tempting to attribute the 0.5°C warming of the past 100 years to the increase in greenhouse gases. Because of the natural variation of temperature, however, such an attribution cannot be made with any degree of confidence."

His critics also objected to what they saw as his style. In referencing his brief stay at the meeting, Schlesinger said, "That is his habit. He comes, gives his talk, and he leaves." This didn't encourage much mutual understanding between Hansen and his peers. A *New York Times Magazine* article later said, "It's not his science that gets Hansen in trouble—it's his style. Hansen has all of the moves of a hustler but none of the guile. Backed by a body of exhaustive and universally respected research, he routinely flouts his profession's tacit restrictions on categorical and unauthorized statements while maintaining the pacific innocence of a curious child."

But there was more beneath the surface than simply disagreeing with Hansen's statistical methods and style. Dr. Stephen Schneider of the National Center for Atmospheric Research said it best: "Hansen is not the villain that people make him out to be. He's a state-of-the-art climate modeler. Hansen got bad press that was partly deserved and partly the envy of other scientists who resent the way he went to Congress." As Stanley Grotch of the Lawrence Livermore National Laboratory said, "If there were a secret ballot at this meeting on the question, most people would say the greenhouse warming is probably there."

Hansen found himself in a Catch-22. Critics said his testimony failed to capture the nuances of the science, but when he had spoken more technically in past testimonies, he couldn't get his point across. Writing in the journal *Science,* journalist Richard Kerr contended, "Had it not been for Hansen and his fame, few in public office, and certainly not the public itself, would have paid much attention to a problem that everyone at Amherst agrees threatens social and economic disruption round the globe." No one had really listened while experts toiled for over a decade on the likely magnitude of the problem. Kerr added, "Then came Hansen. Now greenhouse scientists have the attention they have wanted, but for reasons they think unsound."

Hansen's detractors weren't only in academia. He also faced political opposition from conservative lawmakers who hoped to blunt his message. Hansen did little to alter their political stances. He told us that officials in the George H. W. Bush White House insisted that he make changes to his remarks before subsequent testimony to Congress. When they wouldn't budge, he told them, "OK, go ahead and make the changes. . . . Then I sent a fax to Al Gore [who was leading the hearing] and said, 'Please ask me about this sentence and this sentence, because those were written by the White House." Gore not only agreed, but gave the story to the *New York Times* (with Hansen's permission). The officials were livid, but also reluctant to punish the now famous scientist.

Although his message got the attention he seemingly wanted, oral testimony is not exactly Hansen's favorite thing to do. He said, "I decided after my 1988 testimony that I was going to bail out of this crap and just do science, because I'm a very reluctant speaker. Science is what I like to do."

"But when you find out something really disturbing, you can't keep it to yourself?" we asked.

"Yeah, that's right."

NEW RULES OF ENGAGEMENT

Hansen's withdrawal from the public spotlight lasted the duration of the Clinton administration. At the beginning of the new century, Hansen found himself in hot water again when he squared off with the second President Bush and Vice President Cheney on their energy policy. Hansen identifies as an independent and grimaced at the thought that people might view his climate advocacy as a partisan attack. Nevertheless, he felt the situation was urgent. "I wanted to try to make clear that the Bush administration's energy policy was dangerous." He was further troubled by what he saw was a "gap between what was understood about global warming by the scientific community and what was known about global warming by the people who need to know, the public," a gap that in Hansen's view seemed driven by conservative ideology.

Hansen realized that, under the new Bush administration, NASA had become reluctant to publicize papers that drew attention to climate change. In one instance, he received an e-mail stating, "According to [NASA] HQ, there's a new process that has totally gridlocked all earth science press releases relating to climate or climate change . . . two political appointees . . . and the White House are now reviewing all climate related press releases." Hansen said, "It seemed to me that NASA's Office of Public Affairs had become its Office of Propaganda."

Hansen began to speak out, starting with a speech at his alma mater, the University of Iowa. In December 2005 at the American Geophysical Union in San Francisco, Hansen predicted that 2005 would likely be nearly the hottest on record, a prediction that was featured in the *New York Times* and *Washington Post* and on ABC's *Good Morning America*. In Hansen's word, a "shitstorm" broke at NASA HQ, and the agency's leadership tried to censor Hansen at the behest of the White House. Stricter procedures were immediately implemented by the political appointees in the Office of Public Affairs to prevent him from speaking with the media. NASA removed all global temperature analysis from the Goddard Institute for Space Studies website. They then laid out new "rules of engagement" for all NASA employees, but it was made clear that Hansen was the target.

In response to the "rules of engagement," Hansen said, "I needed to fight back if I wanted to retain an ability to communicate with the outside world." He gave an interview on 60 Minutes, stating, "In my more than three decades in the government, I have never seen anything approaching the degree to which information flow from scientists to the public has been screened and controlled as it is now." He then contacted Andrew Revkin at the New York Times, who wrote an article titled "Climate Expert Says NASA Tried to Silence Him." In response, NASA's inspector general initiated an investigation. Its final report, "Regarding Allegations that NASA Suppressed Climate Change Science and Denied Media Access to Dr. James E. Hansen, a NASA Scientist," confirmed the allegations. In summary, "The preponderance of evidence supported claims of inappropriate political interference in dissemination of NASA scientific results."

So Hansen had quite a track record of fighting to alert the public to what he saw as impending disaster, but was he right? In 1981 Hansen concluded that CO_2-driven atmospheric warming would rise above the noise level of natural climate variability by the end of the century, making humanity an irrefutable dominant force in the climatic system. As NASA data verified, "The 10 warmest years in the 134-year record all have occurred since 2000, with the exception of 1998. More recently, NASA has stated that the year 2015 ranks as the warmest on record."[10] In 1981 Hansen also made a prediction: "The most sophisticated models suggest a mean warming of 2° to 3.5° C for doubling of the CO_2 concentration from 300 to 600 ppm [parts per million]." When he said that, Earth's atmosphere was at 340 ppm. It is now at 403 ppm, and Earth has warmed 0.8°C in the past century.

In explaining his new concern, Hansen again starts in the details. "The upper half of the ocean is gaining heat at a substantial rate," while "the deep ocean is gaining heat at a slower rate, and [this heat] energy is going into the net melting of ice all around the planet, and the land to depths of tens of meters is also warming." For example, in 2012, the total energy imbalance on Earth (i.e., the global energy increase) was about 0.6 watts per square meter. He said, "That may not sound like much, but added up over the whole world, it's enormous. . . . It's about twenty times greater than the rate of energy used by all of humanity, equivalent to exploding four hundred thousand Hiroshima atomic bombs per day, 365 days per year. That's how much extra energy Earth is gaining each day."

What does that have to do with melting ice and sea-level rise? Scientists can analyze Earth's past climate using deep ice and ocean-sediment cores. The technique allows them to look back thousands or millions of years to determine temperature change, CO_2 levels, and relative sea-level change during prehistoric times. They've discovered that Earth's climate history demonstrates a strong correlation among temperature, CO_2, and sea level.

In the absence of any human influence, climate changed in drastic ways, bringing both extreme ice ages with low sea levels and times of little to no ice with sea levels much higher than today. Hansen said, "Small changes in Earth's orbit that occur over tens to hundreds of thousands of years alter the distribution of sunlight on Earth. When there is more sunlight at high latitudes in summer, ice sheets melt. Shrinking ice sheets make the planet darker, so it absorbs more sunlight and becomes warmer. A warmer ocean releases more CO_2, and more CO_2 causes more warming." Thus, even small changes in the climate are, over time, amplified by a self-reinforcing feedback loop, resulting in huge oscillations.

According to Hansen, "The important point is that these same amplifying feedbacks will occur today. The physics does not change. As Earth warms, now because of extra CO_2 we put in the atmosphere, ice will melt, and CO_2 and methane [another powerful greenhouse gas] will be released by the warming ocean and melting permafrost." The question now is how fast these amplifying feedbacks will occur. NASA's own data shows that Greenland and Antarctica are already losing several hundred cubic kilometers of ice each year. "And the rate has accelerated since the measurements began," Hansen noted.

A 2016 study published in *Nature* confirmed Hansen's assertions. This analysis sought to better quantify the rate of West Antarctic ice loss on three glaciers located at the Amundsen Sea. Using data collected by NASA's Operation IceBridge program, scientists were able to determine that, over the course of just seven years, the Smith Glacier has lost 1,000 to 1,600 feet of solid ice from its "grounding zone," the nexus between the glacier and the seabed. In effect, warmer ocean water is carving out the underbelly of the glaciers, leaving them more prone to lose the thinner ice at their edges. At this rate of loss, these glaciers will exist for only three to four more decades, but well before they have completely melted, they are likely to detach from the seabed and float away from Antarctica entirely. These three glaciers, along with the rest of the West Antarctic Ice Sheet, contain enough ice to raise the global sea level by 3.9 feet.

The UN's Intergovernmental Panel on Climate Change (IPCC) is the world's preeminent climate science organization. It is open to all member countries of the United Nations; currently 195 of them are members. Thousands of scientists have contributed to its work. They state, "By endorsing the IPCC reports, governments acknowledge the authority of their scientific content." Therefore, their work is "policy-relevant and yet policy-neutral, never policy-prescriptive." The IPCC has been publishing its climate-change assessments since 1990.

The IPCC disagrees significantly with Hansen about both the rate at which and the level to which the water will rise. The IPCC's most recent assessment was released in 2014. It projects an increase in global temperature at greater than 1°C, but less than about 4°C above preindustrial temperatures. As a result, the IPCC estimates that Greenland's ice sheet could take a millennium or more to disappear completely, bringing a sea level rise of seven meters. It does not estimate Antarctic ice sheet loss because "current evidence and understanding is insufficient."[11] It had not included any projections of global sea-level rise from the melting of ice sheets until this most recent publication.

The IPCC currently projects that by the year 2100, sea levels will rise between 0.28 to 0.98 meters in a relatively gradual, linear increase. The world's governments are basing their efforts to address climate change upon the assumption that sudden ice sheet collapse won't happen, at least not for a long time. Using the IPCC's model, 195 participating countries concurred in the Conference of Parties agreement (COP21), the landmark climate deal reached in Paris in December 2015. It requires participating countries to submit their long-term emissions-reduction targets by 2020, with subsequent updates on their progress every five years.[12] Christiana Figueres, who led the UN climate deal negotiations, said, "One planet, one chance to get it right, and we did it in Paris. We have made history together."

NOT SO FAST

Jim Hansen was not one of the climate scientists applauding the results of COP21. The agreement's warming target is a level he calls "dangerous" in the title of his ice-melt paper. He said, "It's just bullshit for them to say, 'We'll have a 2°C warming target and then try to do a little better every five years.' It's just worthless words. There is no action, just promises."[13] He added in a frustrated tone, "It's doing nothing to change the trajectory that we're on. That's why I call it a fraud. . . . It was OK to get all the countries to say, 'Yeah, there really is a problem and we should solve it,' but to say that you're going to do it by each country trying to do better is a non-solution. Even if some countries do very well, all that will do is keep the price of fossil fuel very low so others can burn it."

Hansen told us that the IPCC's failure to consider ice-sheet dynamics is a huge mistake, the result of trying to make their predictive model look elegant rather than using the data as is. In his words, "What they've fallen into is a trap of competing against each other in a beauty contest in their models." To Hansen, such methods ignore the historical core sampling data, which suggest a potentially dangerous future: the last time the Earth was this warm, there was a massive global flood. "Ice sheet disintegration, unlike ice sheet growth, is a process that can proceed rapidly," Hansen said. "Multiple positive feedbacks accelerate the process once it is underway. These feedbacks occur on and under the ice sheets and in the nearby oceans." The result is a nonlinear, more exponential increase in the rate of ice melt and sea-level rise when global temperatures are pushed beyond a particular threshold.

Why would the IPCC not focus on the effects of melting polar ice? Hansen believes that "scientific reticence," i.e., restraint in coming to a controversial conclusion, is hindering communication with the public about the dangers of global warming. Policy makers need to recognize that, he said. "Scientific reticence may be a consequence of the scientific method. Success in science depends on objective skepticism. Caution, if not reticence, has its merits." He allowed that reticence at the IPCC "is probably a necessary characteristic, given that the IPCC document is produced as a consensus among most nations in the world and represents the views of thousands of scientists." Rather than getting everyone to a consensus with something as complicated as the dynamics of future sea level rise, they may have just punted and said that it's just "too poorly understood."

Nonetheless, Hansen believes we already have a "strong indication that ice sheets will, and are already beginning to, respond in a nonlinear fashion to global warming." Moreover, he warned, "In a case such as ice sheet instability and sea-level rise, there is a danger in excessive caution. We may rue reticence, if it serves to lock in future disasters." In other words, if we wait until we have an irrefutable consensus, it will likely be too late to do anything about the problem.

Hansen believes his latest warning is the most important he has ever issued. Earth's current climate is beginning to look a lot like the way it did during the Eemian interglacial, a warm period from about 130,000 to 115,000 years ago. Then, when the temperature was less than 1°C warmer than today, "there is evidence of ice melt, sea-level rise to 5–9 meters [beyond current levels], and extreme storms."[14] It gets worse. Further evidence of rapid sea-level rise in the late Eemian, "[suggests] the possibility that a critical stability threshold was crossed that caused polar ice sheet collapse."

Ice sheets are vulnerable to warming on multiple fronts both above and below, from warmer air and higher ocean temperatures. The latter is particularly problematic, weakening the ice from below, leading to instability and eventual disintegration and collapse. As a result of all this positive feedback, Hansen's analysis suggests, "sea level is now rising 3.2 mm per year (3.2 m/millennium), an order of magnitude faster than the rate during the prior several thousand years, with . . . Greenland and Antarctica now losing mass at accelerating rates." The volume of ice that is melting is incomprehensible. Recent analysis of satellite gravity measurements found Antarctic mass loss from 2003 to 2013 to be an average of about 67 gigatons per year (one gigaton is one billion tons, or two trillion pounds), accelerating by an average of 11 gigatons each year during that time period. Greenland's estimate was even larger, losing an average of 280 gigatons, accelerating by an average of about 25 gigatons per year.

Let's put the numbers in perspective. Greenland lost 303 gigatons of water in 2014. That's the equivalent of 120 million Olympic-size swimming pools. Placed end-to-end, those pools would go to the moon and back sixteen times.[15] If all of Greenland's ice sheet melted, we'd be in for an additional seven meters of sea level rise. The complete melting of Antarctica's ice sheet would raise the level of the Earth's oceans by sixty meters.[16] Even just losing relatively small portions of Antarctica to warming would have catastrophic consequences for human civilization because many major cities are located on the world's coasts.

Hansen thinks we will likely see a meter of sea-level rise much faster than the IPCC predicts. And once we hit a meter, it's not as though we just stop there and adapt to that given level. It will only accelerate, and it will accelerate quickly. Hansen warns, "Humanity faces near certainty of eventual sea-level rise of at least Eemian proportions, 5–9 meters, if fossil fuel emissions continue on a business-as-usual course."

To simplify the nonlinear aspect of the melting process, Hansen relates it to doubling times. "It will take you several decades to get to one meter, but then to get to two meters takes only a decade. . . . In our paper we say you'll reach multi-meter time scale in 50 to 150 years. But it doesn't really matter in the sense that it will be out of your control and it's going to happen." And, he added, that was only the half of it. There was something called the AMOC, responsible for transporting heat from the tropical Atlantic to the Arctic, and it was about to slow or stop with truly catastrophic results.

The circulation of waters in the North Atlantic near Greenland is the AMOC, or Atlantic meridional overturning circulation, also known as the thermohaline circulation. It is part of a current that flows in a circle around the Atlantic Ocean, including the Gulf Stream, warming northern climates that would otherwise be frigid. The AMOC is the reason why, for example, England is warmer than Quebec, despite being at a higher latitude. Hansen believes that the increasing amount of cold freshwater from the melting ice sheets is "already shutting down the AMOC." This, he emphasized, is a very bad thing. "Shutdown or substantial slowdown of the AMOC . . . will cause a more general increase of severe weather." Because the lower latitudes of the Atlantic will accumulate excess heat, he explained, "It will drive superstorms, stronger than any in modern times. All hell will break loose in the North Atlantic and neighboring lands."[17]

Again Hansen looked to the past to prognosticate the future. He said, "Such a situation occurred in the last interglacial period, 118,000 years ago. The tropics were about 1° warmer than today because Earth's spin axis was tilted less than today. Ocean core data show that the AMOC shut down, the North Atlantic cooled, and there's evidence of powerful superstorms at about that time. [They were] powerful enough for giant waves to toss thousand-ton megaboulders onto the shore in the Bahamas. Some scientists think these boulders may have been tossed by a tsunami, but we present multiple lines of evidence that the boulders and other geologic features are best explained as the result of superstorms."

We are already becoming familiar with these superstorms. Hansen told us, "Observed cooling southeast of Greenland and the extra warming along the United States's East Coast are not natural fluctuations. When the AMOC slows down, it causes both of those. The warm water along the East Coast is the reason that Sandy retained hurricane-force winds all the way up to the New York City area. The nearby Atlantic was about three degrees Celsius warmer than normal. This unusually warm ocean water has also been able to provide moisture for recent record snowstorms." As if we hadn't had enough, Hansen continued. "[AMOC] shutdown then puts you in a situation where it doesn't recover for a few centuries. So it's practically irreversible on any time scale the present public would care about."

We asked him if, as with climate change in the 1980s, the world might eventually come around to see the evidence about sea-level rise and agree with him. He shook his head. "See, this is a really nasty problem because you don't see the lag in the response of the system just because of the inertia of both the ocean and the ice sheets. But then, there's another time constant that comes into play and that's the time constant required for us to change from one energy source to another." Thus, waiting for strong-enough evidence that the ice sheets will collapse may be our undoing. Hansen warned, "If we wait until the real world reveals itself clearly, it may be too late to avoid a sea-level rise of several meters and loss of all coastal cities."

Regarding such inertia and the response of the system, Hansen said, "CO_2 is more recalcitrant than snow and ice, i.e., its response time is longer. CO_2 inserted into the climate system, by humans or plate tectonics, remains in the [system for] 100,000 years before full removal." He added: "This inertia implies that there is additional climate change 'in the pipeline' even without further change of atmospheric composition. Climate system inertia also means that, if large-scale climate change is allowed to occur, it will be exceedingly long-lived, lasting for many centuries."

We looked across the table at each other in silence for a moment, his words hanging in the air. Then he said, "The first-order requirement to stabilize climate is to remove Earth's energy imbalance. . . . If other forcings are unchanged, removing this imbalance requires reducing atmospheric CO_2 from 400 to 350 ppm." He paused and said it in plain English, "I think the conclusion is clear. We are in a position of potentially causing irreparable harm to our children, grandchildren, and future generations."

Meanwhile, Hansen is still receiving the same kind of criticism that he heard in the 1980s. The original version of his recent paper, the one he released for discussion, got more than just discussion. Kevin Trenberth of the National Center for Atmospheric Research, former lead author for the IPCC, said, "The new Hansen et al. study is provocative and intriguing but rife with speculation and 'what if' scenarios. It has many conjectures and huge extrapolations based on quite flimsy evidence . . . it is not a document that can be used for setting policy for anthropogenic climate change, although it pretends to be so. . . . There are too many assumptions and extrapolations for anything to be taken seriously."[18]

Comments from scientists suggested that Hansen is being an alarmist. Sybren Drijfhout and his colleagues of the Royal Netherlands Meteorological Institute said, "The paper and its framing in the public arena sometimes tend to cross the thin line between opinion and scientific evidence. . . . The extreme, abrupt events in this article need some preconditioning and would be much more likely to occur after 2100 or even after 2200. . . . Therefore, we would recommend present[ing] this work with a lower level of 'alarmism,' and avoid terminology [such] as 'dangerous' in the title and upfront displays of the paper."[19] Frustrated, he sarcastically shot back, "Extreme events are much more likely to occur after 2100. . . . Hmm, yes, I guess that we should not be concerned about anything that happens 85 years from now."

Many of his colleagues see him as an "activist" and think that's a bad thing. They think this role has clouded his objectivity as a scientist. Since retiring from NASA in April 2013, after forty-six years, he has participated in many public demonstrations against fossil-fuel exploitation and has been arrested five times.[20] His most notable run-in with the law was when he was arrested outside the White House imploring President Obama with a megaphone to reject the Keystone Pipeline for "for the sake of your children and grandchildren."[21]

Activism allows him to be painted as an alarmist, his warnings prone to being dismissed by mainstream scientists in lieu of more "agreed-upon" science. For instance, Ken Caldeira of the Carnegie Institution for Science said Hansen is "among the best climate scientists," but "it's important to keep value and opinions separate from scientific judgments and empirical fact, and . . . [Hansen] has not made clear enough distinctions." Others think Hansen's effort to affect policy is "diluting" the potency of his views, as stated by Roger Pielke Jr., a policy expert of the University of Colorado.

On whether or not he's an alarmist, Hansen said, "I don't think that I have been alarmist—maybe alarming, but I don't think I'm an alarmist."[22] In justifying his actions, he notes the cost of inaction. "The possibility of such intergenerational injustice is not remote—it is at our doorstep now. We have a planetary climate crisis that requires urgent change to our energy and carbon pathway to avoid dangerous consequences for young people and other life on Earth." When asked about his desire to speak out, Hansen said, "In thinking about whether I was going to speak up or not, what really brought me to this conclusion was, I don't want, in the future, my grandchildren to say, 'Opa understood what was going to happen, but he didn't make it clear.' And so I'm trying to make it clear."

The IPCC reports provide one of the more comprehensive looks into what the future will probably look like with rapid sea-level rise. That institution's assessment is based on its projected level of rise, not

Hansen's. Even though the IPCC's calculations produce more modest flooding than Hansen's, the implications are still catastrophic if we do not adapt. So think of this as the minimum for what to expect:

* Global potential impacts of coastal flood damage and land loss of human settlements are substantial. About 65 percent of the world's cities with populations greater than five million are located in low-lying coastal zones. Such zones constitute 2 percent of the world's land area, but contain at least 10 percent of the global population.

* Many island nations could disappear entirely. Without protection, approximately 187 million people would be displaced due to land loss and submergence by 2100, assuming a sea-level rise of just 0.5 to 2 meters.

* The number of people exposed to devastating annual flooding could reach about 262 million per year given a sea-level rise of 0.6 to 1.3 meters by 2100.

* Vulnerable cities at risk of major increases in annual economic losses include Guangzhou, Miami, New York, New Orleans, Honolulu, Mumbai, Nagoya, Boston, Shenzen, Osaka, Vancouver, Ho Chi Minh City, Kolkata, Guayaquil, and Jakarta.[23] In 2005, the total value of assets exposed to possible flooding from sea-level rise in such coastal cities was about $3 trillion, corresponding to about 5 percent of global GDP.

* By 2070, assuming a rise of just 0.5 meters with respect to today's values, asset exposure grows to at least 9% of global GDP.

* Without global adaptation measures, aggregate losses for coastal cities reaches at least $1 trillion per year for just 0.4 meters of additional sea level by 2050.

It's widely agreed that the sea-level rise predicted in these conservative estimates will occur. But if Hansen is right, it will be much, much worse. A disaster of this scale could collapse the global economy. We asked, "So what you are suggesting is the end of all these coastal cities within a hundred years. If that happens, if countries really can't deal with it, if you can't relocate all those cities, if you can't relocate all those people, is the implication some sort of global systemic economic collapse, bringing about a sort of new dark ages?"

He answered simply, "Yeah," but then he added, "I think I wrote a sentence [in our paper] that the world could become practically ungovernable." Hansen's scenario is one in which massive sea-level rise catalyzes a collapse of human civilization as we know it.

To better understand what the economic effects would look like, Dick tried to find out what it would mean for Boston, where he grew up. There, the city government has already created a sea-level rise committee devoted to addressing this threat. Their maps of Boston at three meters of sea rise show an utterly unrecognizable city. East Boston and Logan Airport would be underwater. Downtown Boston would be submerged. Beacon Hill would be an island. MIT and Harvard, where he studied and later taught, would be gone.

Dr. Charles Fletcher, a professor at the University of Hawaii and leader in the study of sea-level rise, reminded us, "How can a city possibly respond to such a threat? Among the Dutch, who have lived in the shadow of rising waters for centuries, there is a simple, powerful rule of thumb. If you wage war with water, you will lose. Instead, you should yield to water and elevate your cities." Rather than spending billions on dams to keep the water out, the Boston sea-level-rise committee is thinking of letting the water in and creating a system of Venice-like canals in an effort to preserve the city's buildings. Their thinking is that they could turn the system into an economic net-positive, making the canal area of the city into an attractive place to live and visit. But the cost of turning Boston into America's Venice has not been estimated.[24] Even if Boston could afford it, most of the buildings in Boston could not withstand the three to six additional meters of sea water Hansen projects. Then, the coast would move inland, flooding everything in the city and the surrounding metropolitan area.

Throughout the U.S., there have not been detailed plans or cost estimates of the actions that would be necessary to save cities by making them look like Amsterdam or New Orleans, urban areas below sea level protected by dams, levees, and pumps. There is no real understanding of what it would cost to move millions of people and build new cities, no determination of who would pay for it and how, no projection of the taxes required, no understanding of the effect on gross domestic product.

WHAT NEXT?

Despite the pessimism and helplessness we felt listening to Hansen, he was ready to move on to what we should do to stop the disaster. He still thinks it could be stopped with less-than-catastrophic consequences. "You have to actually make the price of fossil fuels reflect their cost to society." He added, "Transition to a post-fossil-fuel world of clean energies will not occur as long as fossil fuels appear to the investor and consumer to be the cheapest energy. Fossil fuels are cheap only because they do not pay their costs to society and receive large direct and indirect subsidies.

"Now, the thing that makes this a tragic situation is that the actions that you need to phase down our dependence on fossil fuels, as rapidly and as practically as possible, actually make economic sense. . . . We actually come down to the fact that it is economically beneficial. . . . If you put a gradually rising fee on carbon, collected from the domestic mines and ports of entry, very easy to do, almost no cost, it's so simple. . . . And if the U.S. and China would do that, the global problem is solved." He added: "A near-global carbon tax might be achieved via a bilateral agreement between [those two countries], the greatest emitters, with a border duty imposed on products from nations without a carbon tax, which would provide a strong incentive for other nations to impose an equivalent carbon tax." The money could then be given back to residents of the country, stimulating the economy and helping to expedite a transition to a clean-energy future.[25]

His idea is to create the incentive the public needs to get behind a more serious push toward renewables, which would make them more economically competitive with fossil fuels. He's bullish that it would raise GNP and create millions of new jobs. "Economically, what makes sense is that you phase out those things that are easiest . . . and people will start paying attention." The carbon fee is essentially the catalyst for a new sustainable-energy revolution to try to prevent Hansen's warning from coming true. He's not sure it will be enough, but he sees it as the best chance we have to avert catastrophe.

"You know, Mark Bowen, who wrote this book about me called *Censoring Science,* said it was remarkable how everything I'm saying now, I actually said three decades earlier" and Bowen believes that new data has only strengthened the case for sea-level rise. "I've shown in several papers over the last five or ten years, look at the rate of change of sea level in any of [my critics'] models and compare that with the real world; it's way too slow. And they sort of ignore that. Nobody refers to [me] because I'm not an expert in ice sheets and such things; they never refer to my papers where I show something like that. But I know it."

Therein lies the distinction between Hansen and his critics, perhaps the key to determining if he is right about future sea levels. They resist incorporating new data into their projections because of its scientific uncertainty, while he is always looking to include new data to come to the "probably right" conclusion in time to act on it.

We think Jim Hansen and his warnings about sea-level rise have a high Cassandra Coefficient. The issue is incredibly complex, without any person, office, or agency clearly assigned to respond to it. In order to believe in rapid ice melt and swift sea-level rise, you have to comprehend the intricacies and relationships of ice movement, ocean currents, and climate science. As Hansen himself points out, even if you do grasp all of that, you still have to make a leap of faith, because the science does not yet fully prove his contention, not in the rock-solid, double-blind way of testing that science employs. Of course, if you wait for that data, Hansen believes, it will be too late. Thus his railing against scientific reticence.

As for the audience, doing something to stop sea-level rise means doing much more than what 195 nations agreed to at the COP21 summit. Taking action to deal with consequences could be far more daunting, moving scores of major metropolitan areas inland and walling off cities with dams and massive pump systems. This is not to mention the unfathomable amount of money required. No government wants to contemplate that task. It is too big, too awful, too disruptive. Sea-level rise is such a slow-moving process that it is very easy for governments to continue to simply do nothing.

Hansen himself seems to exhibit almost all of the characteristics we look for in Cassandras. He is an expert, data driven and science based. He has a track record of legitimate past warnings, not false positives. He has a proven capability to see things clearly, see them first, and be right about them. Hansen assumes a responsibility to do something about what his data shows, even if it means publicly risking his credibility. He has stepped outside of his comfort zone, risking his job, being arrested, working the media. Still, he hopes someone will find something wrong in his calculations, and he seeks peer review in that regard. Even in the face of overwhelmingly negative data, he remains optimistic that humankind can come together to implement a solution. In his case, it is for the adoption of a carbon-tax system that would slow the warming of the atmosphere and the melting of ice.

Back in Midtown after lunch, Dick walked by himself through the throngs of the city, to think through what Hansen had said. He found himself at the Empire State Building and, on a lark, rode to the outdoor observation deck. He looked down on southern Manhattan: Wall Street, the Battery, Little Italy, Chinatown, SoHo, Tribeca, South Street Seaport, hundreds of billions of dollars of buildings and business, the world's economic headquarters. If Hansen is right, it will all disappear below the waves in the lifetimes of people now alive in that city below.

He wanted Hansen to be wrong, but that's precisely the problem. We all want him to be wrong.

James Hansen and Sea Level Rise

CASSANDRA CHARACTERISTICS	THE WARNING	THE DECISION MAKERS	THE CASSANDRA	THE CRITICS
High	X	X	X	X
Moderate				
Low				

CHAPTER 13

The Weatherman:
Nuclear Ice Age

In no other type of warfare does the advantage lie so heavily with the aggressor.

—JAMES FRANCK

A s night fell, a small boat docked near downtown. Ten men got off and climbed up onto the pier. Some hailed taxis. Others just blended into the mass of humanity, the twenty million people who crowded into the seaside city. Hours later it began, a three-day orgy of death and destruction that put one of the world's megacities on lockdown. Eight thousand miles away, on the campus of Rutgers University in New Jersey, the massacre left American climate scientist Alan Robock fearing his theory was about to be tested.

The men from the boat had taken automatic weapons and bombs from their backpacks and assaulted nine facilities, including the main train station, the two finest hotels, and a Jewish cultural center. They had never been to Mumbai before, but they had studied their targets using Google images and maps. They kept in touch with their controller on mobile phones. The controller watched the Indian police and army response on television and advised his attackers how to adjust their plans.

The Indian security services responded poorly. The counterassault command team trained to deal with such an attack was hundreds of miles away in New Delhi. It took many long hours to get an airplane to fly them to Mumbai. After the smoke cleared, after the fires in the hotels were put out, after the scores of bodies were transported to the morgue, RAW, as the Indian intelligence service is known, finally figured out where the attackers had come from. The "Research and Analysis Wing" analysts, with some help from their American counterparts, had also determined the answer to where the attackers' controller had been during the assault. The answer to both questions was the same: Pakistan.

One of the ten attackers was captured alive. Under interrogation, he also confirmed that he had come from Pakistan, where he had been trained by a group known as Lashkar-e-Taiba, a terrorist unit with links to the Pakistani military's interservice intelligence organization, known as ISI. In response to this information, India went on military alert and began to move forces toward the Pakistani border.[1] The Indian parliament and public demanded revenge. It was 2008, and President George W. Bush was warned by his advisors that his last year in office was about to include a war between two of the largest and most capable militaries in the world. In an effort to prevent such an occurrence, he dispatched his national security advisor and the most senior U.S. military officer to New Delhi to try to talk the Indians down and try to get the Pakistanis to cooperate in an investigation.[2] It was a dangerous moment.

The Pakistanis detained the leaders of Lashkar-e-Taiba, briefly placing them under house arrest. The fever eventually passed in New Delhi, though Indian citizens, officials, and lawmakers alike were still experiencing anxiety, fearing another attack was imminent. The Indian leadership assumed it would happen again; after all, Lashkar-e-Taiba had attacked the Indian parliament in New Delhi in 2001, the equivalent of a terrorist attack on the U.S. Capitol. Indian security services had concluded in that case, too, that Pakistani military intelligence, ISI, had been behind the terrorism. India's military had also gone on full alert then, giving the world a war scare that took months to pass. The problem for New Delhi was that it had taken three weeks to go on full alert and move troops to the Pakistani border in 2001. That was too long.

After the attack on the parliament in 2001 and especially after the Mumbai terrorist attacks in 2008, the Indian military decided that in the future, it needed to be able to respond to Pakistani terrorism before diplomats could intervene to calm things down. The Indian military saw this cycle of terrorist attacks followed by panicked diplomacy as failing to address the fundamental problem. For India's strategists, the issue was Pakistan's apparent conclusion that it had effectively subdued and tamed India (despite India's larger and more advanced military) by possessing nuclear weapons. The Pakistanis believed that no Indian leader would ever authorize Indian military retaliation against Pakistan, fearing that any retaliation would trigger a nuclear war. Having created a nuclear deterrent that prevented conventional war between the two subcontinent rivals, Pakistan's military now felt free to sponsor terrorist groups to put pressure on India over the disputed ownership of Kashmir.

For the Indian military and civilian leadership, that situation called for a change. Indian officials felt they had to call Pakistan's bluff. India had to be able, once again, to go to war against Pakistan, and had to do so in a manner that did not provoke a nuclear escalation. If Pakistan engaged in another major terrorist attack, the Indian

Army must be able to punish Pakistan and do so quickly. It had to be capable of initiating combat within seventy two hours or less. They called this new doctrine Cold Start.

Under Cold Start, if there were another Pakistani-sponsored terrorist attack against India, there would be enough Indian Army units near the Pakistani border to initiate combat within three days. The Indian military ensured that these units, more than any others, would be the best trained, equipped, and supplied. Those units would have more than enough fuel and ammunition. They would each have predetermined plans and objectives in Pakistan. They would all be supported by Indian Air Force aircraft, which were already moved to air bases within range of Pakistan. The Indian Navy would put enough submarines and other firepower on the Arabian Sea to find and sink the Pakistani Navy ships on patrol and to keep the remainder of the Pakistani Navy bottled up in its few harbors. To pay for Cold Start, the Indian parliament increased defense spending by 50 percent.

Based on the three wars that India and Pakistan have fought since the two nations came into existence in 1948, it is not hard to imagine the degree of aggression with which India would attack. Supported by close air support and artillery, India's armored corps would roll into the Pakistani Punjab and toward the major city of Lahore, just across the border in central Pakistan. In the south, Indian forces would speed across the salt plains of the Rann of Kutch and the Thar Desert toward Pakistan's megacity of Karachi on the Arabian Sea. The Indian attack would stop, however, perhaps fifty miles into Pakistan. Cold Start would have limited objectives.

The objective would be to damage Pakistan's military and its economy and also to demonstrate that India was not deterred by Pakistan's nuclear weapons. India would attempt to show that it could still use its military strength to force Pakistan to pay a price for its terrorism in India. Perhaps Indian strategists might think that in the wake of a limited military defeat, Pakistan's politicians would take control of their discredited military and intelligence services and reign them in. As part of that thinking, Cold Start is designed to have a "Fast Stop." India would achieve its limited objectives quickly and would then announce a cease fire, so as not to put the Pakistani military up against a wall, so as not to threaten the sovereignty of Pakistan. An all-out military invasion, Indian strategists realize, could provoke a nuclear response from Pakistan. Thus, India has designed Cold Start to not be overly provocative.

The problem with Cold Start, American analysts fear, is that it might underestimate Pakistan's willingness to use nuclear weapons. The Indians seem to think that Pakistan would resort to nuclear weapons only in an existential situation, one in which Pakistan was about to be eliminated as a functioning state, or its military totally vanquished and destroyed. Some evidence suggests that India could be wrong about that. There is also reason to worry about exactly who would make the decision to launch Pakistani nuclear weapons.

Pakistan is a nation of almost two hundred million people. It emerged from the British colony of India in 1948 when, following Britain's departure and decolonization, Muslim leaders wanted their own nation, separate from the majority Hindu nation of India. The resulting partition was violent, with widespread fighting and millions of Hindus and Muslims moving from one nation to the other. In the decades since the two nations came into existence, there have been wars or threats of war between them. Frequently, the controversy surrounds the contested ownership of the Jammu and Kashmir regions in the north. Each nation's military and intelligence service see the other's as their raisons d'être and their biggest threat.

India, however, has more than six times the population of Pakistan. Indeed, within India today there are almost as many Muslims as there are in all of Pakistan. Its armed forces, at 1.2 million on active duty, are twice the size of Pakistan's. It is in large measure to deal with this asymmetry that Pakistan developed its own nuclear weapons, beginning in the early 1970s. The Pakistani activity had the effect of increasing the pressure on India to divert resources to its own nuclear weapons program. Though both countries were hard at work, each trying to make their own operational bomb, India beat Pakistan to the punch by nearly a decade. India detonated its first nuclear weapon in 1974 (Operation Smiling Buddha), while it was not until 1983 that Pakistan was able to begin subcritical testing of its weapons. The year 1998 marked the most testing by both nations, a total of eleven different devices detonated, to the great dismay of the signatories of the Comprehensive Test Ban Treaty which had been adopted by the UN only two years prior. Since then, experts agree, Pakistan has manufactured and deployed more nuclear weapons than its larger neighbor. Both nations now have nuclear weapons in a variety of configurations, both bombs and missile warheads.

A REGIONAL TINDERBOX

The Pakistani Army has watched over the last decade as the size and capability of Indian Army units near the border has increased. It has also observed a great improvement in Indian logistics, with more fuel, ammunition, and trucks to carry them moving closer to the units on the border. These buildups are similar to things that the Soviet Red Army did in the 1970s and '80s in Germany, and also to what the North Korean Army has done in the not-so-distant past. Both of those armies sought to minimize the time it took from the decision to start fighting to when the first attacks could begin. They developed the concept of the "standing start" and reduced the number of activities that the other side could see them doing, activities that would indicate seriousness about going to war. This was largely achieved by simply focusing on war preparation in advance, in steady-state peacetime.

Pakistan has since responded by doing some similar things, like reducing the time it takes to respond to an attack, or as the American military puts it, the time from flash to bang. When both sides start making it easier to go to war quickly, they are effectively creating a nuclear hair trigger by eliminating the time when cooler heads could prevail and destroying the window during which commanders can verify conditions to ensure that there has not been a mistake. When it's all about seeing how fast you can mount an attack, there's a real potential that you'll misfire, be offside, or pull the trigger too fast. Once again, the American military has a saying that captures the essence of the current relationship between India and Pakistan: "ready, fire, aim."

In addition to logistical improvements, Pakistan has added a new kind of weapon for its army, tactical nuclear weapons. These are low-yield nuclear warheads on short-range missiles or artillery shells designed for battlefield use. Until recently, both India and Pakistan had only strategic nuclear weapons, high-yield bombs and long-range ballistic missile warheads designed to be dropped deep in the enemy's territory, on command headquarters, airfields, missile bases, logistic hubs, and even political and economic targets. These tactical warheads are intended to be mounted on missiles with a range of less than forty miles. Given that these missiles' launchers are likely to be somewhat behind the front lines, they will penetrate only a few miles into enemy-held territory, blowing up attacking ground forces just far enough away from Pakistan's own front lines so as not to damage its own troops. The tactical nuclear weapons were designed to blunt an Indian Cold Start penetration of Pakistani territory, to stop the invading forces cold.

The U.S. and the Soviet Union used to have such weapons, the most notorious of which was the Davy Crockett, a jeep-mounted nuclear artillery shell that was a last-resort weapon designed to stop a Soviet breakthrough. Thousands of American and Soviet tactical nuclear weapons were deployed a few miles apart in Germany for thirty years. However, most tactical nuclear weapons have now been retired from the U.S. arsenal.

On its face, tactical nuclear weapons do not sound like a terrible thing, because they're low yield and can't level an entire city. Therefore some may believe that they are not as dangerous as a strategic nuclear bomb, but tactical nuclear weapons introduce a different problem. Because they're low yield, equivalent to "only" a few hundred tons of TNT, there could be a diminished psychological reluctance to use them.

Use of a low-yield weapon could easily trigger an escalating response. For example, a shell with a yield of a hundred tons could provoke a retaliatory strike with a one-kiloton shell, which could provoke a counter-retaliatory strike with a ten-kiloton warhead, and so on. Tactical nuclear weapons also emit nuclear fallout after detonation (though not as much as a large warhead), and can still expose anyone who wanders into the fallout zone to harmful levels of ionizing radiation for years to come.

Pakistan's nuclear doctrine has apparently evolved from a policy of strategic use in response to a threat to its military or sovereignty to a policy of use early in a war to blunt a quick and limited Indian punitive retaliation against a terrorist attack. Pakistan now might use tactical nuclear weapons to prevent an Indian breakthrough, and to prove to India that it is, in fact, willing to use its nuclear arsenal in a fight.

A U.S. raid in 2011 prompted Pakistan to change where its tactical nuclear weapons are stored, how they're assembled, and who can order their use. Shortly after midnight on May 2 of that year, U.S. military forces penetrated Pakistani air space and inserted commandos into what the Pakistanis call a garrison city, a city controlled by its army. In this case it was the home of the War College, the city of Abbottabad. Neither the U.S. Blackhawk helicopters nor the U.S. fighter-bombers showed up on Pakistani radar. The nation's air defense system had not even been alerted to any intrusion of its airspace whatsoever. The Americans were in and out before the Pakistani military had even realized that the Americans had paid them a visit.

It was a great embarrassment for the Pakistani generals. The Pakistanis were not much troubled by the fact that Osama bin Laden had been living down the street from their War College for a few years (though some would argue that many of them may already have known that). No, the most frustrating thing to the Pakistani brass was the fact that the United States could fly a commando team in stealth helicopters into one of their military towns, conduct a raid, and get out before Pakistan could detect it, much less respond. If the Americans could do that to a terrorist compound, could they do it to a nuclear weapons storage facility? Clearly the U.S. had been flying stealthy reconnaissance drones over Abbottabad for weeks without being detected. Had they also been checking out the nuclear sites? After all, there had been rumors for years that the U.S. had thought about seizing, stealing, or destroying Pakistan's nuclear weapons. Now, as the senior military staff met in an unusual emergency session in May 2011, they realized that the Americans had the ability to pull off such an operation if they wanted. Forget Osama. Worry about the nukes. That was the theme all the generals could agree on.

As a result, according to some accounts, the military began dispersing its nuclear weapons from a handful of well-guarded bunkers to a much larger number of bases. It also reportedly initiated a program to frequently and covertly move the weapons to new locations, with only a handful of officers knowing where all the weapons were at any given time. Rather than moving them around in easily detected,

heavily guarded convoys, they allegedly started putting them in une-
scorted, inconspicuous trucks.[3] This concept is called security by obscurity, and it works well as long as nobody figures out which lightly
guarded truck contains a nuclear weapon. Should a terrorist group
learn of a particular truck with a bomb, perhaps from a sympathetic
insider (and there are known to be terrorist sympathizers in the Pakistani military), it would be easy for such a group to outgun and seize
such an insufficiently protected vehicle.

Historical policy had called for the nuclear weapons to be stored
in pieces, which were to be assembled into complete weapons only
when conditions presented themselves that could conceivably warrant their use. This "separation of parts" policy was designed so that
even if some terrorist or rogue commander seized a weapon, the device would not work because an essential component was missing.
Now, reports state, because Pakistani military doctrine might call for
tactical nuclear weapons to be used on short notice, some weapons
would be kept fully assembled.

These reports raise the possibility that a local Pakistani military
commander might have the ability, if not the authority, to launch a
nuclear weapon to save his unit from being overwhelmed by an Indian
Cold Start attack. Likewise, a terrorist group might be able to seize a
fully assembled weapon, as opposed to just a component of one. In
May 2011, a terrorist group attacked and penetrated a Pakistani Navy
base near Karachi, a base widely believed to contain nuclear weapons. Four other bases alleged to have nuclear weapons have also been
penetrated by terrorist attacks between 2007 and 2013. The frequency
of these events does not instill confidence in the security of Pakistani
nuclear warheads.

Pakistan also significantly stepped up its rate of production of
nuclear weapons, according to media accounts. Some experts have
estimated that Pakistan will, within a few years, double its stockpile of

nuclear weapons from two hundred to four hundred. One thing that most experts agree on is that Pakistan is producing nuclear weapons faster than any other nation on Earth. Four hundred warheads would mean that Pakistan's nuclear inventory would surpass not only India's, but also France's, the United Kingdom's, and China's. This would mean that Pakistan would have fewer than only the United States and Russia.

The result of Cold Start, the bin Laden raid, and the Pakistani response to those events is that the probability of a nuclear exchange on the Asian subcontinent has greatly increased. India has developed a quick-response capability to another Pakistani-sponsored terrorist attack, premised on an educated guess of where Pakistan's threshold for nuclear weapons use is. It stands to reason that Pakistan has subsequently lowered that threshold by way of the aforementioned policy changes and increased weapons production. Pakistan may have made it easier for a besieged commander in wartime, or a rogue commander in peacetime, to unilaterally initiate a nuclear strike against India.

American academics who were nuclear-war theoreticians were shocked in the early 1960s to learn that the United States had only one nuclear war plan: try to go first, but in any event, fire everything we had as quickly as we could. In reaction to learning of this "strategy," American bureaucrats, as well as U.S. military tacticians, engaged in a significant effort in the 1970s to design "limited nuclear options" or LNOs. The idea was that actual use of nuclear weapons, even just a few, would cause the leaders of the United States and the Soviet Union to be so frightened of escalation that they would stop, back off, and de-escalate. Even in graduate school at MIT, Dick spent endless hours learning about, and doing the math behind, LNOs. The problem was that we were never sure that the Soviets had embraced the idea of limited nuclear war. Some post–Cold War research suggests that the Soviet military never implemented LNO concepts and would not have stopped once nuclear weapons were used, until they had carried out their full strike plans, or until they had launched their full arsenal.

The reality is that no one really knows what will happen when one side starts using nuclear weapons against another side that is also nuclear armed. It is possible that escalation control will be unachievable in the chaos of nuclear war. On the Asian subcontinent, it could be that any nuclear use will escalate and continue until both sides run out of nuclear weapons. That could easily happen during only one day of rapid nuclear exchanges.

The Indian and Pakistani nuclear inventories together may now total fewer than five hundred weapons, but these inventories are climbing rapidly. That is indeed a small number compared to the U.S. and Russian nuclear inventories, which once totaled over twenty thousand nuclear weapons each. Nonetheless, if the two nations used several hundred weapons on each other, there would be millions of what nuclear warriors call "prompt deaths," followed by millions more as a result of radiation sickness or exposure over the following years. One estimate is that the "prompt deaths" would total twenty million people, or roughly forty percent of all the deaths in World War II. The death toll could also be much higher, given that the two countries currently have a combined population of 1.4 billion people. That estimate assumes that Indian and Pakistani cities would be targeted for the sake of taking out military and political command facilities, logistics systems, airfields, and ports.

The Asian subcontinent would be hell on Earth for any of those 1.4 billion people who survived all of the nuclear strikes. Famine would occur in major swaths of both countries. Radiation sickness would be widespread. The medical system would break down. The economies of the two countries would totally collapse. Given the interconnectedness of the global economy, we could see a global financial reaction of unprecedented severity. Ultimately, it would be a calamity on a scale seldom seen in world history. Alan Robock thinks it could actually be far worse.

THE NUCLEAR WINTER

Robock is a weatherman. His diplomas, obtained at a technical institute in Cambridge, Massachusetts, in the 1970s, were in meteorology. In the 1980s, he started thinking about global weather and what could adversely affect it in major ways. He examined the effects of large volcanic eruptions, drawing on historical records. Meanwhile, during this same decade, a well-known astronomer, Carl Sagan, became the public face of a group of scientists who modeled what would happen to global weather patterns if there were a nuclear war between the United States and the Soviet Union. The group became known as TTAPS, after the last names of the scientists: Richard P. Turco, Owen Toon, Thomas P. Ackerman, James B. Pollack, and Carl Sagan.

TTAPS said in 1982 that a nuclear war between the United States and the Soviet Union, something that seemed possible at the time, would do far more than ruin those two countries and their allies in Europe (and possibly Asia). The scientists said that it could mean the end of humanity altogether. In 1983, Sagan pulled together a group of twenty-eight scientists, including both Americans and Russians, for the Conference on the Long-Term Worldwide Biological Consequences of Nuclear War, held in Washington in 1983. The results of the meeting were published in 1984 in *The Cold and the Dark: The World after Nuclear War*. The book received widespread attention in the media and, as we later learned, at the highest levels of government in Washington and Moscow.

In retrospect it seems inexplicable that almost forty years after the first nuclear weapon was used, after the two superpowers had built tens of thousands of nuclear weapons, that the world's most powerful nations did not really understand the implications of using them against one another. After all, they had come perilously close to fighting such a war several times, the Cuban Missile Crisis of 1962 being the most famous. Both nations had nuclear-armed missiles on alert, capable of launching at a moment's notice. On a normal day, the United States had B-52 bombers armed with nuclear weapons flying patrols just outside of Russian airspace, each with predetermined targets, waiting for the "go code." Despite that massive adoption of nuclear weapons, there were things about them that we didn't initially understand. One was the electromagnetic pulse, or EMP. It was only after several nuclear tests near Las Vegas "fried" electronics many miles from the test site that scientists began to realize that electromagnetic waves emitted during a nuclear explosion could travel well beyond the blast radius and destroy the circuitry of electronic devices. Another thing that they had not understood, said Sagan and company, was the global climatic effect of numerous, simultaneous, large scale firestorms caused by a nuclear war.

Although many of us have now internalized the idea that major thermonuclear war between two superpowers could mean the end of the world, the leaders of the United States and the Soviet Union at that time, indeed the leaders of most countries, did not believe that. Both the U.S. and the USSR had plans, however fanciful, of surviving and reconstructing after a nuclear war. Both nations spent enormous resources on protective shelters and programs to be able to rebuild if a nuclear exchange ever took place. There were probably people in places like Brazil and India who thought that in a post-nuclear-war world their nations would become the new superpowers of the planet. However, Sagan and company said that there would be no surviving and reconstructing. There would be no replacement superpowers.

Because of the climatic effects of numerous large firestorms, there would be the gradual death of most, if not all, humans. It was a chilling prediction, and it came from scientists, from experts. It became known as the nuclear winter theory.

The theory, although it involves complex computer modeling of the global climate, can be easily explained. First, it assumes that nuclear war commanders will target cities. American strategists called this a countervalue strike, a euphemism, as contrasted with a counterforce strike, which meant hitting each other's nuclear forces and military infrastructure.

Second, it adduces the fact, known in principle from tests and directly from the Japanese experience, that a nuclear explosion in a city creates a firestorm, a raging fire of such intensity that its updraft sucks a powerful, high-speed wind into the fire at ground level, sustaining the combustion. The high temperatures ignite virtually everything, moving and spreading the blaze throughout the entire city. The extreme heat from the fire dries out other combustible material, igniting other things and spreading the fire even farther. Allied bombing using incendiary devices had caused such firestorms during World War II in Hamburg, Dresden, Nagoya, Yokohama, Tokyo, Osaka, and Kobe and, using nuclear weapons, in Hiroshima and Nagasaki. Moreover, an all-out nuclear war would involve multiple detonations over many major cities of weapons far more powerful than the Hiroshima bomb.

Third, the TTAPS group explained that such fires, consuming buildings and everything in them, create a type of soot that is projected into the upper atmosphere by the updraft. An immense amount of such soot would be blown upward as firestorms, aided by gas mains and oil tanks ruptured by the blast, completed the obliteration of dozens or hundreds of cities.

This soot, they claimed as their fourth assumption, would remain aloft, because it would be heated by sunlight and would be above the rain-forming cloud layer. It would disperse with global wind patterns, eventually covering both the Northern and Southern hemispheres.

Their fifth assumption was the most devastating of them all. The soot would block so much sunlight that temperatures would fall and organisms that rely on photosynthesis would die, i.e., plants that sustain much of the life on Earth. In addition, the ozone layer would be damaged, allowing more ultraviolet radiation and cosmic rays into Earth's biosphere, further destroying crops and damaging DNA in most organisms. There were complex formulas and models, but the bottom line was that a major nuclear war would mean that the survivors would envy the dead and would likely join them a lot sooner than they had expected. The group disagreed among themselves whether the result would be the complete end of humanity as a life-form on Earth, or merely the end of human civilization as we know it today.

Some scientists disagreed, claiming that there were flaws in TTAPS's calculations, or at least that they relied on unproven assumptions. Sagan responded that it was impossible fully to test the hypothesis, as science normally demanded, without having a nuclear war. Anti–nuclear weapons groups in Europe and the United States had been seeking a "nuclear freeze," a ban on the creation of any additional nuclear weapons. The nuclear-winter theory gave these activists additional impetus, and their movement became a significant political force on both sides of the Atlantic.

Sagan, along with TTAPS partner Richard Turco, reflected on their cause in another book, *A Path Where No Man Thought: Nuclear Winter and the End of the Arms Race*, arguing for further nuclear weapons reductions. In it, they recall the story of Cassandra as told in Aeschylus's play *Agamemnon*: "Nobody paid attention. . . . They didn't want to hear. . . . Today she would be dismissed as a prophet of doom and gloom . . . she can't understand how it is that (her) predictions of catastrophe—some of which, if believed, could be prevented—are being ignored." They summarize what the leaders of ancient Troy told Cassandra: "We can't be sure, and if we can't be sure, we are going to ignore it."

Reflecting upon Cassandra's fate, Sagan and Turco noted that "the resistance to dire prophesy that Cassandra endured is equally stubborn today." They tried to imagine why disaster predictions are still ignored. "Mitigating or circumventing the danger might take time, effort, money, courage. It might require us to alter the priorities of our lives." Therefore, the two scientists submit that the typical reaction is: "Improbable. Doom and Gloom. We've never experienced anything remotely like it. Trying to frighten everyone. Bad for public morale." The authors suggest that leaders tend to have a "natural, if someone maladaptive, tendency to reject the whole business. It will need much better evidence. . . . There is a tendency to minimize, dismiss, forget," perhaps because they feel that they are being blamed for having allowed the condition to develop. Sagan and Turco could well have written the book that you are now reading, for their observations still hold true for a wide range of possible Cassandras, like Alan Robock.

ROBOCK'S WARNING

It is hard to know precisely what the political effects of the nuclear-winter theory were, but those who were the leaders of the U.S. and USSR at the time later admitted that it had helped prompt them to act. Ronald Reagan was the U.S. President at the time and was widely thought to be eager for a fight with the Soviet Union. Yet after reading about the nuclear-winter theory, Reagan met with his Soviet counterpart, Mikhail Gorbachev, in Iceland and proposed the abolition of nuclear weapons. The Russian leader, who also had a growing concern about a nuclear winter, agreed in principle, but suggested that the two countries start by limiting the deployment of new weapons and then later reducing their stockpiles.

In the years that followed, interest in the nuclear winter theory diminished. The Cold War ended peacefully. The Soviet Union and its military alliance, the Warsaw Pact, dissolved. Thousands of nuclear weapons were dismantled and destroyed in the U.S. and the former USSR. Thousands more nuclear weapons were withdrawn from Western and Central Europe. Some scientists, reflecting on the nuclear winter theory, suggested that it had been exaggerated for political effect. The world moved on. Scientists began to focus on ozone depletion and the greenhouse-gas phenomenon, which came to be known as global warming and then climate change.

Alan Robock remained concerned. In 2007, he recalculated the effects of a nuclear war, using vastly more sophisticated global climatic, atmospheric, and ocean data and simulations than had been available in the 1980s. He confirmed the nuclear winter theory. The effects of nuclear war had not been exaggerated in the slightest, but, in fact, were worse than originally predicted. A nuclear winter would require less soot ejected into the upper atmosphere, and the effects would be longer lasting than had earlier been thought to be the case. Thus, Robock argued, even though the United States and Russia had far fewer nuclear warheads than they possessed decades earlier, they still had more than enough to create a global nuclear winter.

In 2009, with Owen Toon, he looked at the possibility that some similar effect might occur even without the involvement of either the U.S. or Russia. In their article in *Scientific American*, "Local Nuclear War, Global Suffering," they examined a scenario in which India and Pakistan had a nuclear exchange. They modeled their results on the assumption that each side used only fifty nuclear weapons (of Hiroshima size, quite small by today's standards) on countervalue targets, i.e., cities. Their model suggested that the flammable cities would create five million tons of soot and smoke that would quickly find its home in the upper atmosphere.

The climatic and ecological effects from such a "regional war" would be global, they found, caused by the dispersion of soot in the mesosphere and the further depletion of ozone in the stratosphere. Temperatures would drop precipitously. In the middle of the continents, places like Kansas, it would be below zero in the summer. These conditions would be largely comparable to the "Little Ice Age" of the sixteenth to nineteenth centuries, colder than any time in the last two thousand years or more. Using new agricultural models, they predicted that many crops would fail, including 40 percent of China's winter wheat, 35 percent of China's spring wheat, and 20 percent of U.S. corn. Famine would be widespread, all from the use of fewer than one tenth of 1 percent of the world's nuclear weapons.

The reaction to Robock and Toon's work was, not surprisingly, minimal. Robock refined and double-checked the calculations and published in peer-reviewed journals, *Science* and *Nature*. He went to a broader audience, writing in the Huffington Post and learning how to use Twitter. He grabbed the President's science advisor, John Holdren, on the margins of a meeting. He wrote an op-ed that appeared in the *New York Times*. Fidel Castro's office called and asked for a briefing, so Robock went to Havana, but he was not asked to brief the U.S. President. He was not asked to explain his work in New Delhi or Islamabad.

Robock is a distinguished meteorologist and climate scientist. He is not a political scientist, foreign-policy expert, arms-control analyst, or national-security consultant. When it comes to figuring out how to reduce the danger of the risk he has reaffirmed, he thinks in terms of the bigger picture. "We should eliminate all nuclear weapons, all of them. They all have to go," he said. It's the same conclusion that

Ronald Reagan came to decades ago. Robock doesn't try to come up with some localized solution to minimize the risk of war in India and Pakistan. "All of them, all nuclear weapons. They all have to go." Some national-security, foreign-policy, and arms-control experts agree. None other than former secretaries of state George Shultz and Henry Kissinger have also called for the elimination of all nuclear weapons. Their call, too, has been largely ignored.

In his mid-sixties, Alan Robock has thought of retiring, but he feels he has to go on. He has to get the message out that the threat of a nuclear winter has not gone away. Sitting with Robock, Dick could feel his frustration. It was palpable. "I would love for someone to show me our calculations are wrong." He sipped his coffee. "I mean, we published in peer-reviewed journals. Other people have replicated the results with their models."[4] He does admit that there are variables that could make the impact slightly less awful, if fewer cities were struck, if there were massive rainstorms at the time of the war, if they're wrong about the size and strength of firestorms in modern subcontinent cities. "We need funding for grants to study the oceanic and atmospheric models," he says, but well-studied large-scale volcanic eruptions, especially Tambora in 1815, Krakatau in 1883, and Pinatubo in 1991, have already shown that one localized ejection of soot into the atmosphere can have global climatic effects.

Reflecting on the lackluster reaction to his work, Robock told us, "People don't want to hear it. It's too big of a problem. They want to walk away and assume somebody else will deal with it. It's depressing." Echoes of Sagan channeling Cassandra bounced around the room, but we still had to ask why *he* didn't just walk away and leave it to others. What made him feel that it was his job to reduce the danger? "I take the public's money in research grants. If I discover a danger to society, it's my duty as a scientist to warn the public about it."

Although Alan Robock never got into the Oval Office to give his briefing, he might be surprised to find that the man sitting there in 2016 agreed with his goal. Barack Obama, at a summit of world leaders focused on the nuclear threat, stated:

> I said in Prague that achieving the security and peace of a world without nuclear weapons will not happen quickly, perhaps not in my lifetime, but we have begun. As the only nation ever to use nuclear weapons, the United States has a moral obligation to continue to lead the way in eliminating them. Still, no one nation can realize this vision alone. It must be the work of the world. We're clear-eyed about the high hurdles ahead, but I believe that we must never resign ourselves to the fatalism that the spread of nuclear weapons is inevitable. Even as we deal with the realities of the world as it is, we must continue to strive for our vision of the world as it ought to be.

Robock would no doubt urge the U.S. President to "lead by example by eliminating all but maybe two hundred U.S. nuclear weapons to start things rolling." Dick asked if that would persuade India and Pakistan to eliminate all of theirs. "I would hope so," Robock replied.

Alan Robock and his warning of the possibility of a regional nuclear war triggering global nuclear winter have a high Cassandra Coefficient. The issue is complex. It requires an extensive understanding of political-military affairs, South Asian politics, nuclear weapons, agronomy, and climate science. Doing something about the risk means doing something about the seventy-year-old enmity between India and Pakistan or, even more unlikely, convincing the world's nine nuclear-weapons states to give up what they see as the ultimate guarantor of their security.

There is no one clear set of possible responders. Is the audience the Indian and Pakistani leadership, just as it was Reagan and Gorbachev in the 1980s? Or is it the broader world community, which seems much less focused on "banning the bomb" than it did in the 1960s, '70s, and '80s? Scientists, although mostly antiwar and anti–nuclear weapons advocates, seem to quibble with the data and want more proof.

Robock is clearly a proven climate scientist, an renowned expert. He is data driven, but still wishes he had more precise data. He seeks comments, feedback, and criticism from his peers, hoping they can find errors that will prove him wrong. Until they do, however, he feels a strong sense of obligation, a duty that goes beyond himself. He does what he knows best. He writes journal articles to get the word out, yet he remains caught up in a seemingly everlasting personal struggle to uncover what more he can do to sound the alarm about the threat of a nuclear winter.

Alan Robock and Nuclear Winter

CASSANDRA CHARACTERISTICS	THE WARNING	THE DECISION MAKERS	THE CASSANDRA	THE CRITICS
High		X	X	
Moderate	X			X
Low				

CHAPTER 14

The Engineer:
The Internet of Everything

It's the great irony of our Information Age—the very technologies that empower us to create and to build also empower those who would disrupt and destroy.

—PRESIDENT BARACK OBAMA

'Twas the night before Christmas, 2015, and Dick was planning to smoke a fresh hunk of venison for the next day's dinner party. He instinctively picked up his iPhone to check his e-mail. That was a mistake. Most people had stopped their usual chatter to settle in for the holidays, but not Joe Weiss. The e-mail he sent could've had the subject line "I told you it would happen." Instead, it was simply "Ukrainian electric grid attacked." After reading his message, Dick put aside the venison for a bit and spent the next hour on the Web verifying what Weiss had written. As usual, it seemed he was right. Weeks later, U.S. intelligence and the Department of Homeland Security came to the same conclusion.

The small Ukrainian city of a quarter-million people in the foothills of the Carpathian Mountains looks like a mixture of *The*

Sound of Music and *Doctor Zhivago*. Although in Ukraine, it is just a short drive from Hungary, Slovakia, Poland, or Romania. Its streets and squares are dotted with Armenian, Greek, Ukrainian, and Roman Catholic churches. Stanislau or, as it has been officially known for the last fifty years, Ivano-Frankivsk, is an unlikely ground zero for a dangerous emerging technological threat. The prior evening, the Christmas lights had shut off throughout the entire city. In fact, all the lights went out, as did all of the electricity.

It was Prykarpattyaoblenergo, the state's electric utility, that had the problem. Someone had hacked into the control consoles in its operations room, then accessed the controls for thirty electric substations and thrown the breakers, similar to throwing the breakers on your home's electric power panel. The result was the same: the power went off. Neighborhoods served by the affected substations were plunged into darkness. It was only four thirty in the afternoon, but the winter sun had just set and the temperature was dropping fast. Some residents found their telephones in the dark and tried to call the power company. Few could get through. A deluge of calls was flooding the lines at Prykarpattyaoblenergo, but most of the flood were robocalls originating outside of Ukraine.

The operations room staff was horrified to discover that, despite a backup generator, their computers had also gone dark. The consoles were frozen, and the software had been wiped clean from the hard drives. Unable to reset the breakers remotely, the utility personnel put on their winter jackets and headed for the parking lot. Without computerized controls, the staff was forced to drive out to each substation, unlock the fences, find the switches, and manually turn them back on. That time-consuming process was the only way to reconnect the electricity from the main grid to the substations and then to the houses, apartment blocks, and commercial establishments that were now in the dark. Shortly after ten that evening, the utility company had the power back to normal, but most of their computer screens were still showing the "blue screen of death." They had been killed by something called BlackEnergy.

BlackEnergy is not so much a single piece of software as it is a family of malware. When it was first detected in 2014, it seemed to be a program that generated a flood of requests on the Internet, a digital denial of service, overwhelming a targeted server with traffic to prevent most genuine messages from getting through. However, further computer forensic work showed that BlackEnergy was also being used to steal data and to maintain backdoor access to a user's system. A new feature was later added, called KillDisk, a "wiper" program that removes all software from a targeted hard drive, turning a computer into little more than a paperweight. BlackEnergy is believed to have been developed by a shadowy group known as Sandworm, as in the creatures from the Dune science-fiction books. The malware family seems not to have been developed on that desert planet, however, but somewhere in Russia, according to forensics experts. Why would Russians want to throw a nice little Carpathian city into darkness during the Christmas season? Retaliation.

Things had not been going well between the countries that once were the two largest republics in the Soviet Union. Russia and Ukraine had been in a slow-motion war since Russia's President Putin had ordered the invasion and occupation of Ukraine's Crimean Peninsula in 2014. Then, in late November 2015, alleged "Ukrainian nationalists" threw Russian-occupied Crimea into a power blackout. They did it the old-fashioned way, blowing up the power-transmission towers with explosives. Crimea, whose electricity came mostly from Ukraine, was plunged into weeks of darkness. Almost exactly one month after the first attacks on the Crimean power grid, the Russians retaliated against substations in Ukraine. This time the attackers did it the new-fashioned way, with software.

For almost two decades, Joe Weiss had been saying that could happen, yet American electric power companies have refused to believe him. Off the record, their spokesmen call Joe an alarmist, an exaggerator, a spreader of FUD (fear, uncertainty, and doubt).

Joe Weiss is a short, balding, bespectacled engineer in his sixties. He has been involved in engineering and automation for four decades, including fifteen years at the respected Electric Power Research Institute. He has enough initials after his name to be a member of the House of Lords (PE, CISM, CRISC, IEEE Senior Fellow, ISA Fellow, etc.), all of which speak of his expertise and qualifications as an engineer. He holds patents for things like the steam turbine fuzzy logic controller. He wrote the safety standards for automation of nuclear power plants. He also wrote the book on the issue we are discussing here, *Protecting Industrial Control Systems from Electronic Threats*, not a best seller but the definitive work in the field.

Today he runs his own consulting practice, which he operates from his home in the upper-middle-class neighborhood of Cupertino, California, not far from Apple Headquarters. Dick first met him in 2002 when he asked the U.S. Commerce Department to host a conference on the possibility of cyber threats to the power grid and other industrial controls. Weiss stole the show. His passion, as well as his expertise, showed through his precise, engineering language. Nevertheless, representatives of electric power companies and the industry's self-regulatory body, the NERC (North American Electrical Reliability Corporation) instructed Dick during and after the conference not to pay too much attention to Weiss. He was a zealot, an outlier, excitable. He overstated the problem.

But what if he were right, Dick wondered. Weiss's ultimate contention was that it was possible to hack your way from the Internet into electric power company enterprise networks and from there to the control systems that ran the grid. Subsequent tests performed by the federal government showed that that was almost always true. Weiss was right. Under pressure from the federal government and the threat of regulation, the NERC emerged from its state of denial and gradually adopted cybersecurity regulations of its own. Weiss was not satisfied.

"The utilities got together and came up with a set of criteria, called the NERC [North American Electric Reliability Corporation] critical infrastructure protection [CIP] standards. In those standards they input a number of exclusions and allowed them to self-define what would be 'critical.' Telecommunications they excluded. Can you imagine doing a cyber assessment of your IT systems and being told, 'Do not address telecom'?" Weiss fumes. "It simply doesn't make any sense." The exclusions, he contends, the things the industry doesn't want regulated, are precisely the areas that need standards. "NERC said almost 70 percent of power plants in U.S. were not considered critical. Almost 30 percent of transmission assets were not considered critical. All of the distribution assets, which are the heart of the smart grid, are not considered critical because distribution is explicitly excluded."[1]

Even after the Ukrainian blackout, the electric power industry continued to disagree with Weiss about the significance of that event and the vulnerability of the U.S. grid. Kimberly Mielcarek, a spokeswoman for the Electricity Information Sharing and Analysis Center (E-ISAC), flatly rejected that it could happen here. "There is no credible evidence that the incident could affect North American grid operations and [there are] no plans to modify existing regulations or guidance based on this incident," she said in a statement sent to Reuters news agency. Weiss vehemently disagrees. "Where is the uproar from such an egregious statement?" he asked. What was attacked in Ukraine, he explained, was a part of the grid classified as "low voltage transmission and distribution systems." The U.S. electric industry standards for cyber security do not even cover that part of the grid, Weiss pointed out. "What happened in Ukraine could certainly happen here."[2]

Indeed, in November 2014, the federal government officially informed the power-grid companies about the threat to them from the BlackEnergy malware and the activity of supposed Russian entities targeting U.S. power companies. The Department of Homeland Security gave briefings to power companies, the contents of which are still not publicly available. Later, according to Weiss and other sources, Russian entities electronically stole data regarding electrical breakers and their controls from several U.S. power companies. Why the breakers are so important, according to Weiss, is not just because they can disrupt the power for a few hours, as was done in Ukraine. If done in a particular manner, manipulating the breakers can also result in a blackout that lasts for weeks by crippling and destroying critical equipment that is in short supply and takes a great deal of time to replace.

To explain how, Weiss says we need to look at another small foothills city. Idaho Falls claims a "metro" population of 140,000 and is best known for the National Laboratory on its outskirts, a place with numerous nuclear reactors. On the grounds of that laboratory in 2007, a team of U.S. government employees hacked their way into the controls of an electrical grid operations room and took remote control of breakers as part of a cybersecurity exercise. The similarity to what happened eight years later in Ukraine is striking. The difference, Weiss says, is that the 2015 attackers stopped after only the first step. They could have, but didn't make the next move, which would have destroyed key equipment on the grid. Had they continued, the electrical infrastructure for Idaho Falls would have experienced what is known as an Aurora event after the code-name for the test.

As Weiss explains it, the problem is "remote hacking of variable frequency drives of large, rotating equipment in power plants, refineries, dams," that is, hacking into big things that spin, things like generators. All generators connected to a power grid must spin at exactly the same speed, even if they are hundreds of miles apart from one another. This is referred to as the electricity being "in phase." If they are not spinning at the right rate, the result, Weiss says, can be "devastating." This is because the grid briefly turns generators into electric motors if they are out of phase with each other. The generator undergoes a tremendous amount of mechanical stress in the few microseconds that the grid is forcing its rotation either faster or slower.

When engineers start a generator, it must first be disconnected from the grid until it is spinning at the same rate as all the others on the grid. Once its spin rate is verified, the engineers can flip the breakers to reconnect the generator to the grid. Otherwise, what happens is what occurred during the 2007 Idaho Falls exercise. The hackers threw the breakers early and the generator began to malfunction. The exercise was halted before pieces of metal began flying through the air.

Weiss explains: "Opening breakers is step one in creating an Aurora event. The second part is simply reclosing the breakers out of phase with the grid. If you can remotely open the breaker, you can remotely close it. The Ukrainian outage could have been so much worse if the attackers had chosen to do so. Considering most U.S. utilities have still not installed Aurora hardware mitigation, it just may be a matter of time before really bad things happen." When automatic control systems on generators do not do their job, Weiss notes, large pieces of metal can indeed fly through the air.

EVERYTHING IS CONNECTED

SSH is what experts on electrical control systems call the gigantic hydroelectric complex in Siberia known as Sayano-Shushenskaya

Hydro. The largest hydro plant in Russia, its construction was as important to the 1960s Communist government as the Hoover Dam was to Roosevelt-era America. It was a monument to the ability of the state to tame rivers, generate electricity, and build on a monumental scale. Its ten generators housed in the Great Hall below the immense dam were an integral part of this national treasure.

On August 16, 2009, a nearby hydroelectric plant at Bratsk shut down because of a fire. The grid operations center, the Siberian Unified Dispatch Control Center (SUDCC), shifted the lead for the Automated Load Frequency Control System for all Siberian spinning generators from Bratsk to SSH and specifically to the newly upgraded Turbine 2 (T2). The recent improvements to T2 had included new software, "a new automatic control system meant to slow or speed up the turbine to match" the grid.

The operators at SSH did not make that decision to shift the grid load to themselves, nor did they control the turbine. The turbines were monitored and controlled remotely from the SUDCC. T2, now connected to and leading all of the other generators in the Siberian grid, began to vibrate. By morning of the next day, bolts buckled. Then, with a loud pop, the cover securing the turbine shot through the roof, collapsing the ceiling of the Great Hall. Unrestrained, T2, a 1,500-ton blade fan, rose fifty feet in the air and then, still spinning, flew through the Great Hall, shredding equipment and killing seventy-five people. The automatic water cutoff system failed, and the flow that was supposed to be passing through the generator at high speed became a geyser shooting sixty-seven thousand gallons of water a second into the Great Hall. The flood short-circuited some of the other generators, which then exploded, causing at least forty metric tons of oil to cascade out into the river, which then killed four hundred tons of trout in two nearby fisheries.

When the rescue teams arrived the next day, the Great Hall below the dam looked like the site of a nuclear detonation. The before-and-after photographs are startling. Giant pieces of equipment were shredded and submerged. Walls, ceilings, everything was in disarray and barely recognizable. Floating in the rancid water were seventy-five corpses.

No, Joe Weiss admits, the SSH disaster did not result from a hack, but it does demonstrate that generators spinning at the wrong rate and connected to the grid can fail catastrophically. Throw the breakers at the wrong time and you get a giant generator hurtling across the room, destroying everything in its path. SSH took five years to repair.

"If the worst thing you could manage with a cyber attack on the grid were a two-or three-day outage, that would really, in the grand scheme of things, not be that significant," Weiss says. However, physically destroying equipment that is not kept in inventory, equipment that is custom-built only when ordered, is another story. SSH is an accurate depiction of the magnitude of destruction that could be caused by a hacker with nefarious intentions.

Generators are not the only things that spin. So do centrifuges that enrich uranium for nuclear power plants. "The recent Ukrainian attack was clearly specifically orchestrated and targeted, just like Natanz," Weiss told us. A heavily guarded, underground facility in Natanz, Iran, had eight hundred centrifuges for the country's nuclear program. One day they started to spin at abnormal speeds, sometimes too fast, sometimes too slow. The wear and tear destroyed them from the inside out. Yet all the while, the sensors that should have reported the variations in spin rate to the control room showed that all equipment was performing nominally. Thus did the United States slow the Iranian nuclear weapons program for a time, using a hack, a piece of malware known as Stuxnet.

Why Natanz is important, says Weiss, is that it showed how corrupting digital control system software allows a hacker to send the wrong signals to a programmable logic controller (PLC), the computer inside machines that controls what that machine does and how it does it. Digital control system software packages are running millions of PLCs throughout the U.S. infrastructure, not just in the power grid, but also in pipelines, refineries, and manufacturing facilities. Also known as industrial control systems or supervisory control and data acquisition (SCADA) systems, these software packages are ubiquitous. Versions of them now run our automobiles and control the medical devices in hospitals. And these SCADA systems are networked, sending diagnostic data to and receiving software updates from the manufacturer. "This is about a lot more than just hacking the power grid," Weiss told us. The same SCADA software used by the Iranians is used in thousands of U.S. manufacturing and operating plants and facilities and is susceptible to the same exploits that the U.S. used to destroy those Iranian centrifuges.

This computer network, transparent to most of us, is a critical part of the global infrastructure. It is known as the Internet of Things, the IoT. "The Internet of Things is just a marketing term that somebody thought up long after millions of machines were already networked," Weiss explained to us. "And most of them are networked in ways that can be accessed, perhaps indirectly, from the public Internet." It is that understanding, that everything is connected, that keeps Joe Weiss up at night.

How many devices are connected? The consensus estimate is that by the end of this decade, worldwide, there will be about fifty billion connected machines comprising the IoT. Initially, the computer industry talked about the phenomenon less grandly as simple M2M, machine-to-machine, or D2D, device-to-device, communication. The idea was simple enough and has been around for decades. Rather than have some poor fellow standing out in the cold for hours to turn a crank to stop oil from flowing down a pipeline or to flip a switch after a train passed, you could send an electronic signal down a line and cause the crank to spin or the switch to move.

The IoT is not something that will happen. It has already happened, and it is expanding rapidly. Already sensors on elevators, electric motors, manufacturing robots, and aircraft are phoning home, reporting on their condition and asking for further instructions. Are they in good working order? Are they decaying faster than was anticipated? Are they overheating? Are they running out of consumables like printer ink or cans of Coke? Most of the IoT consists of one machine reporting to another machine and that second device dispatching a human to fix the problem. What made these systems vulnerable was the widespread adoption of digital, rather than analog, signaling, meaning that all machines began speaking in the same "language." Compounding this was the use of the Internet to communicate with equipment from an operations room that could be located anywhere in the world. Next, Weiss points to San Bruno.

Not far from San Francisco International Airport, San Bruno is a middle class residential suburb, not an industrial town. Yet under the ground in San Bruno was a gas pipeline, controlled by SCADA software that used the Internet as its communications backbone. On September 9, 2010, a short circuit caused the operations room to see a valve as open when it had actually closed, spiking the readings coming from pipeline pressure sensors in different parts of the system. Unbeknownst to the families returning home from ballet and soccer practice, technicians were frantically trying to isolate and fix the problem. At 6:11 p.m., a corroded segment of pipe ruptured in a gas-fueled fireball. The resulting explosion ripped apart the neighborhood, most of whose residents had no idea they lived near a pipeline. Eight people died. Seventeen homes burned down. The utility, PG&E, was hit with a $1.6 billion fine.

The accident investigation report blamed the disaster on a substandard segment of pipe and technical errors. There was no suggestion that the software error was intentional, no indication that malicious actors were involved. "But that's just the point," Joe Weiss argues. "The Internet of Things introduces new vulnerabilities even without malicious actors." The problem, Weiss claims, is using Internet software for things that it was never intended to run, like industrial controls, or linking solid industrial control software over Internet communications networks.

The icon of the IoT and the darling of Silicon Valley techies and entrepreneurs is a round, wall-mounted gadget called Nest. Invented by two former Apple engineers, Nest was bought by Google for $3.2 billion in cash. Essentially a thermostat connected to the Internet, Nest also has software that learns your behavior and adjusts the temperature in your house on its own. It also checks on the Internet for the weather in your zip code. More recent models of Nest are linked to door locks, lights, window shades, and cameras. Unlike most of the IoT, which is hidden from consumers inside of machines, whirring in factories and office buildings, Nest occupies a prominent place on the wall of the home. Indeed, it controls the home. It has become quite popular within the consumer electronics industry in recent years.

However, in mid-January 2016, there was a little problem with a software update from the month before. *New York Times* reporter Nick Bilton described his personal experience: "The Nest Learning Thermostat is dead to me, literally. Last week, my once-beloved 'smart' thermostat suffered from a mysterious software bug that drained its battery and sent our home into a chill in the middle of the night. Although I had set the thermostat to 70 degrees overnight, my wife and I were woken by a cold and crying baby at 4 a.m." His Nest had died, its battery depleted by the software glitch. Thousands of other Nest users also woke in the cold because when the Nest died, it shut off the heat. Other users complained that their home alarm systems had triggered in the middle of the night for no apparent reason, ripping them from a condition of deep sleep into a state of high panic. Nest apologized and suggested that users perform a complex nine-step process to re-vivify their little round home-control systems.

One market where the Internet of Things purveyors are pressing for widespread adoption is medical care. One estimate is that the IoT-medical market could reach $117 billion by the end of the decade. Among the device types that are or will soon be netted into the IoT are pacemakers, implantable cardiac defibrillators, insulin pumps, IV pumps (including those for prescription painkillers), and heart-lung machines. The advantages seem obvious. These devices could detect your medical problem and perhaps administer treatment, even if you and your doctor are miles apart.

The problem is that the health care industry, which is rushing headlong into the IoT, has a bad track record when it comes to cyber security. BitSight, a Boston firm that ranks companies for their level of cybersecurity, compared five industries: health care, finance, retail, utilities, and federal agencies. Health care, represented by 2,500 companies in the survey, placed dead last. Veracode, another Boston cyber company (for which Dick serves as a member of the board) also looked at five industries, but with a different metric. It asked what percentage of known vulnerabilities in software had been fixed. In manufacturing, over 80 percent of the problems had been addressed. In medicine, it was half that number. In fact, more than three quarters of all medical software applications currently in use have a known vulnerability. Often medical devices are running on software that is so old that its creators no longer try to fix it, akin to an old version of Microsoft Windows from the 1990s. Even when software companies do send out patches to fix newly discovered vulnerabilities, some medical devices cannot process them if their hardware is too old.

Medical devices and other existing machines being tied into the IoT do not have processors and memory that can act quickly enough to handle even very simple security measures, Weiss says. "Simply performing a virus definition update on an older control-system processor can cause anywhere from a two-to six-minute denial of service. That's just doing your daily virus definition update!" As a result, Weiss thinks that the first step in securing the IoT is building entirely new devices with faster processors and more memory. In essence, hundreds of billions of dollars' worth of machines need to be replaced or upgraded significantly.

Joe Weiss's concerns were confirmed again in October 2016, when over one hundred thousand video surveillance cameras connected to the Internet were compromised by a hacking group, turned into a botnet, and then used to launch a flood of Internet traffic, a distributed denial of service attack, on one of the largest Internet address lookup sites, Domain Name System servers. The result was that hundreds of large Internet sites were unavailable until the attackers turned off the flood. The cameras lacked a firewall to prevent someone from placing malicious software on them. Often, devices like those cameras do require a password to access them, but equally often the manufacturer writes its own password in the source code so that it can always get in. Once this so-called hard coded password becomes known on the Internet black market, the devices are easily enslaved for malicious use. With billions of such devices going online in the latter half of this decade, the potential number of bots is enough to disrupt the Internet itself, or any site or network on it.

UNIVERSAL HACKABILITY

Weiss is not alone in his concern. Academic experts have been worried for years. For example, Kevin Fu at the University of Massachusetts noted that, at one Boston-area hospital, 664 pieces of medical equipment are running on older Windows operating systems and are essentially unprotected from the malicious software that floats around on the Internet. "I find this mind-boggling," Fu says. "Conventional malware is rampant in hospitals because of medical devices using un-patched operating systems." The chief information officer of that Boston hospital admitted that he worried about "situations where blood gas analyzers, compounders, radiology equipment, nuclear-medical delivery systems, could become compromised to [the point that] they can't be used, or they become compromised to the point where their values are adjusted without the software knowing."

One way that medical devices could be compromised is by a botnet that thinks they're just personal computers. A botnet is a prepro-grammed, "fire-and-forget" piece of malware that just knocks on every door on the Internet, every Internet Protocol (IP) address which each and every device connected to the Internet has. If a botnet knocks and can get in, it usually sets up shop without the knowledge of the computer's owner. Then it begins to do its mission, like sending out spam e-mails or participating in a DDoS attack. Medical devices connected to the Internet also have IP addresses, and there is not necessarily anything that tells the botnet that they are medical devices. Thus, a botnet could infect an IV pump and use it to spew spam around the world, monopolizing the processor while degrading the functioning of the pump.

Despite repeated warnings from Weiss and others, nothing seems to slow industry's mad dash toward connecting everything. The latest arena is automobiles. Long before the self-driving car craze, cars were already becoming part of the IoT. The academic expert we know in that arena is Stefan Savage, who hangs his hat at the beautiful campus of the University of California at San Diego. He famously demonstrated to a conference how to hack a Corvette. As *Wired* reported, Stefan used "a 2-inch-square gadget that's designed to be plugged into cars' and trucks' dashboards and used by insurance firms and trucking fleets to monitor vehicles' location, speed and efficiency. By sending carefully crafted SMS messages to one of those cheap dongles connected to the dashboard of a Corvette," Stefan took control of the 'Vette.

The attacks on Corvettes are not just theoretical or even unintentional. In October 2015, for example, highway patrol officers in Riverside, California, were able to locate and recover a stolen Corvette by using its Internet connection to track the vehicle, slow it down, and then kill the engine. As a Corvette owner, Dick isn't sure how he feels about this connectivity. If his Stingray is stolen, he is certainly happy that it is connected, but if some nutjob sends his car crashing into an eighteen-wheeler on the interstate, he will probably think otherwise. Corvettes are not uniquely vulnerable. In fact, it's difficult to find a car these days that is not part of the IoT.

At one level, Joe Weiss has succeeded. Once he was a lone voice in the wilderness about the need for cybersecurity for connected machines, especially on the power grid, but today it is accepted wisdom that devices can be hacked and that even the power grid can be a target. In October 2015, a popular CBS television series, *Madame Secretary*, aired an episode in which the American President retaliates against Russia for hacking Air Force One by ordering the U.S. military to turn the power off in Moscow. We suspect few American viewers turned to their companions to ask, "Can we really do that?" Gradually, over the last decade, universal hackability has become an accepted part of our consciousness.

Is he a Cassandra? a reporter asked. He dodged the question. "There are people who are starting to believe the control system cyber [threat] is real, but it is still a small fraction. All I have are facts and physics[, but] there is an unfortunate tendency to simply want to declare victory."

Weiss is not declaring victory. Indeed, he continues to sound the alarm. The fact that some people recognize there is a problem and a few of them have started to do something about it is cold comfort.

He's concerned that the overall problem has not been solved and is actually getting worse because of the rapid growth of the IoT. His recommendations are specific: a massive research-and-development effort to create inherently secure industrial control systems and devices, machines that can easily have their software updated when vulnerabilities are discovered, rapidly authenticate who is communicating with them, and quickly encrypt and decrypt signals. Some devices, he believes, need to communicate over secure networks that cannot be communicated with from the publicly accessible Internet.

We came away with the belief that Joe Weiss and warnings against connecting everything warrant a high Cassandra Coefficient. The issue is difficult to articulate and, therefore, difficult to understand. Devices with Internet connectivity are coming online in breathtaking numbers, but in a largely unseen manner. Complexity is growing in the very machines that control the underpinnings of modern societies, nearly all of which are being developed and deployed with substantial cyber vulnerabilities. Finally, there are numerous shadowy, malevolent actors learning how to take advantage of this growing set of vulnerabilities. The solutions to this massive challenge are neither obvious nor singular. There is no silver bullet, and correcting these vulnerabilities will be a drawn-out, expensive endeavor.

The audience for Weiss's warning is a diffuse one. Should the responsible party consist mainly of government regulators, and if so, which ones? Is it the responsibility of software vendors or the companies that use their industrial control systems, or of the software engineering and computer science community? Regulators are under political pressure not to regulate. Corporations are under market pressure not to spend money on things that are not required. Software developers are incentivized to get products to market quickly and cheaply. Computer scientists complete undergraduate and graduate programs often without any formal coursework in cybersecurity.

Weiss himself is clearly an expert, having spent thirty years in his field. He is data driven, establishing and maintaining the most complete database of cyber-related industrial control-system incidents. Not content to make the case just to his consulting clients, he has taken on the personal mission of sounding the alarm that significant damage can and will be done because of the lack of security in industrial control systems and on the Internet of Things. The word we most often heard from others in his field when talking about Weiss was *obsessed*. Some said it in a way that implied that it was a bad thing. It is, however, almost a sine qua non of Cassandras.

Is it going to take a "Cyber Pearl Harbor" to learn if Joe Weiss was indeed a Cassandra, someone who was right about predicting a disaster but whose advice was not acted upon sufficiently? For Joe, that future bad day is unnecessary to establish his bona fides. "I have 750 incidents in my database now that have killed over a thousand people [in all]. . . . There have been cyber incidents in almost all of the industrial infrastructures: electric distribution system, transmission systems, hydro plants, fossil plants, nuclear, combustion-turbine plants, oil and gas pipelines, water and water treatment systems, manufacturing facilities, and transportation. These are worldwide, not just in the U.S. None of this is hypothetical. It has already occurred."

Joe Weiss and the Internet of Things

CASSANDRA CHARACTERISTICS	THE WARNING	THE DECISION MAKERS	THE CASSANDRA	THE CRITICS
High	X	X	X	
Moderate				X
Low				

CHAPTER 15

The Planetary Defender: Meteor Strike

It rained iron to the East of Lake Erh-hai. Houses and hill tops were damaged with holes. Most of the people and animals struck by them were killed. It was known as "the iron rain."

—CHINESE GOVERNMENT RECORD, C. 1321

A massive airburst explosion erupted without warning above a Soviet air defense base in Siberia, throwing fighter aircraft around, igniting fires in fuel supplies, killing hundreds, and injuring many more. The Soviet military command in Moscow, unable to determine what happened, thought it could have been a U.S. sneak attack. They assumed that more attacks were minutes away and considered launching a nuclear response against America. Gradually, however, it became apparent that the explosion had been an asteroid hitting the Earth, creating an effect similar to a nuclear detonation in the atmosphere, but without the radiation. That is the story that astrophysicist David Morrison told in the opening pages of his seminal

1988 book *Cosmic Catastrophes*, coauthored with Clark Chapman. The story was fiction, sort of.

No Soviet air-defense base has been ripped apart by a meteor strike, but everything else in Morrison's detailed account has actually happened. A massive airburst explosion from an asteroid took place above the Tunguska region of Siberia, destroying everything within two thousand square kilometers (772 square miles) of forest. It did happen, but in June 1908, about an hour after sunrise.

Millions of people in northern Europe saw a bright light flash across the sky. The few people in Siberia, mainly non–Russian speaking indigenous reindeer herders, had also seen the burst. People miles from the scene were knocked to the ground by a shock wave from the blast. Some, farther away, were startled by a deafening sound. In addition to the thousands of humans who felt the earth shake that day, the recently installed, first-ever, global seismic measurement system felt and recorded the event as well. Oddly, no one immediately went to look at the land under the point of detonation. At least no scientists went. Some herdsmen may have cautiously approached the area, but there is no record of such a journey. Perhaps this is because the epicenter of the meteor strike, the point directly below where the detonation had taken place, was a three-day trek by reindeer sled through some of the densest forest in the world.

Not until 1929, over twenty years later, did a scientific team make it all the way to the Tunguska site. The Russian scientists were both awed and disappointed. They were amazed at the millions of felled trees, snapped backward in a circular pattern away from the epicenter, all singed and charred, but not burned to ash. What left them crestfallen was the absence of an impact crater. They had been hoping to find a giant meteor, which they might have taken back to their laboratories in Leningrad, but there was no hole in the ground, no giant rock. In that era, sixteen years before the first human-caused nuclear explosion, thirty years before human-manufactured space-ships first reentered the Earth's atmosphere, little was known about atmospheric blast effects or the physics of atmospheric reentry. By the 1980s, with extensive test data to supplement what had been recorded over five decades earlier, scientists and engineers were able to calculate what happened that last night of June in 1908, and it terrified them.

The Siberian blast was, they believe, caused by an asteroid ninety meters in diameter striking the Earth's thick atmosphere at high speed and quickly superheated by the friction of its descent, heated to the point of explosion about twenty thousand feet above the surface. At that height, lower than today's transcontinental flights, the asteroid vaporized because of overheating and, in so doing, let loose a blast wave equivalent to five to fifteen megatons of TNT. Even at the lower end of that range, it would have been over three hundred times larger than the nuclear bomb dropped on Hiroshima. The downward pressure from the blast completely devastated eight hundred square miles, about twenty times the size of Washington, D.C. It was one of the largest explosions on Earth ever recorded by humans.

David Morrison, in his book, had simply shown what would have happened if the 1908 event had taken place at the time of his writing. It might have triggered global thermonuclear war. If the asteroid had been a few minutes later in arriving at Earth, its impact could have been over Helsinki or Oslo. Even in 1908, that would have killed hundreds of thousands of people. The message of Morrison's book was that the 1908 Tunguska event was not as anomalous as we would like to think, and a strike of equal or greater magnitude could happen again.

Morrison, who was born just before World War II, is an astronomer with a Ph.D. from Harvard and decades of experience leading space-exploration projects at NASA. He is also generally regarded as the person who focused the world's attention on the threat from asteroids and is a leader in the search for extraterrestrial intelligence (SETI). In *Cosmic Catastrophes*, he slowly made the case for worrying about asteroid impact in the way a good prosecutor would carefully and convincingly lay out his evidence to a jury. He began off-planet.

For over three centuries, humans had been looking at the moon through telescopes. What they saw was a great pockmarked expanse, a surface covered in craters. For centuries astronomers had assumed that the Moon had gone through a tumultuous period of volcanic explosions, causing the craters, before settling down to being the dead rock we now know. Morrison explained how the scientific community had eventually abandoned that theory for the realization that the craters had been formed by meteor impacts. Then, when humans sent cameras to other planets and moons, they also discovered craters, evidence that our neighboring planets had been repeatedly hit by meteors as well. How had the Earth escaped this bombardment?

The answer that Morrison and other astronomers and geologists came to was simple. The Earth had not escaped. Earth is, religious myths notwithstanding, 4.2 billion years old. The oldest things we have found that one might consider human attempts to record history are cave drawings that are perhaps forty thousand years old. The oldest human writing samples are less than six thousand years old. If large asteroid impacts were to occur on average every ten thousand years, there might not be a human-made record of one. Indeed, even during most of the few millennia of human writing, the population was spread so thinly over the planet, most of which is covered by water, that an asteroid impact might not have been noticed by anyone capable of recording it. Even as recently as 1908, the explosion in Siberia almost went unnoticed. It took two decades for a scientific team to make it to the epicenter. A few hundred years from now, there will hardly be any evidence even at the scene, since the trees will have grown back. Larger impacts, however, should have produced craters. Why are there no craters?

Morrison's answer is clear: there *are* craters. In his case for worrying about asteroids and meteors, he begins his exposition of Earth's cratered surface by mentioning Arizona, specifically the city of Winslow. The year Dick graduated from college, the rock band the Eagles had a hit song called "Take It Easy," in which they recalled "standing on a corner in Winslow, Arizona, such a fine sight to see." For most Americans, that song is the extent of their knowledge about this small town. The year before that song came out, Dick had been to Winslow as part of the then de rigueur rite of passage, a cross-country road trip with college buddies. After leaving Winslow, they turned off the highway onto a side road to see a crater that was created by an asteroid impact long ago. When they walked up to the rim and looked down, they were startled. It was massive, far larger than they had anticipated. Almost four thousand feet across and five hundred feet deep, it looked like the surface of the Moon, on which humans had landed for the first time just two summers before. Indeed, the humans who had recently gone to the Moon had first come here, to Winslow, to train on moonlike conditions in the crater.

The Arizona impact occurred an estimated fifty thousand years ago. The first humans arrived in the area about forty thousand years after the explosion, but not until 1906 did a scientist think it was an impact crater. After all, there was no meteor in it. People had assumed it was a volcanic remnant. The scientific community didn't really adopt the impact-crater theory until 1960, so it was still a newly accepted idea when Dick drove up to the giant hole in 1971. Like so many tourists who came to the site, he asked, "So where's the meteor?" There were little bits of meteorite material around, but no big rock. Like the thing that had flown at Siberia in 1908, it had been vaporized by the intense heat created by the friction of its entry through the atmosphere. The crater was caused not by the impact of a meteor digging itself into the surface, but by the blast wave from the atmospheric explosion.

Craters visible to the naked eye, like the one near Winslow, are not a common sight around the world. There are only a handful of large impact craters on our planet, the aforementioned crater in Winslow, another in Australia, one in Brazil, and a few more. Would that mean that the Earth was special, that it had escaped the bombardment that had happened on other planets and moons? Astronomer and geologist Gene Shoemaker studied the Arizona crater in the 1960s and began a study of Earth's craters in general. His answer was yes, the Earth is special, but it had indeed been bombarded by asteroids all the same. Some of the reasons the Earth is special is that it has a thick atmosphere, it has a vibrant biological and geological ecosystem, and it has weather. The atmosphere causes many incoming asteroids to explode in the air rather than on the surface, significantly reducing the chances of an impact crater. The active ecosystem removes evidence of craters by rounding them off and filling them in over centuries of weathering and vegetation. That simply doesn't happen on the Moon or Mars, or the other rocky planets we have seen through telescopes and spacecraft.

After focusing his attention downward at the Earth's craters, Shoemaker turned his gaze upward and began surveying the skies for possible incoming asteroids. Almost thirty years later, he spotted a comet and plotted its course. He predicted a direct impact with devastating results, and he was proven right. His prediction, that the comet would hit Jupiter, happened on schedule and was caught on camera, the first time humans had recorded a major impact on another planet. Despite the cloudy, gaseous conditions on Jupiter, the impact left a large scar.

MOTHER NATURE'S COVER UP

When NASA scientists began studying Earth from spacecraft at the urging of people like Shoemaker, David Morrison, and Jim Hansen (the same Hansen we met in chapter 12), they found thousands of "touched-up" craters, some of them enormous. Some of these meteor craters are obvious, such as the ones in Texas, Utah, and Indiana, but others are so large and weathered that the thousands of people living in villages in them are unaware of the crater's existence. Morrison used a Bavarian crater as an example. The Reis Crater is fifteen million years old, measures fifteen miles across, and is dotted with quaint German towns. When it was explosively created, the effects were felt across the planet. Now you could traverse across the entire crater and never notice that any of that ever happened. Mother Earth is a cosmetic genius.

Perhaps her greatest cover-up is of Chicxulub, the largest asteroid impact on Earth of which we are now aware. It happened sixty-six million years ago, but was not clearly recognized until forty years ago, and the implications were not clearly understood until about ten years ago. The belief that the dinosaurs were wiped out by an asteroid was not a widely accepted scientific theory until relatively recently. It was first argued persuasively in a scientific journal in 1981, but the location of the impact was not known. That same year, geologists who had been searching for oil in Mexico released data suggesting that a massive impact crater existed hidden under the surface of the Yucatán Peninsula. Sixty-six million years of weathering and continental drift had completely covered it up.[1]

David Morrison credits these twin discoveries in 1980, of the Yucatán impact and the subsequent mass dinosaur extinction, as leading to a broad consensus in the scientific community that the one had caused the other. He says that was when he began to think seriously about the effects of asteroid impacts. "Very few people had thought about it before then," he said.[2] "Thirty years ago there was no research on near-earth asteroids. There weren't many known and hardly anything to study." Nonetheless, he points out that "the asteroids were always coming close" and some were entering the atmosphere, "but no one noticed them." Since then, the U.S. Air Force has placed satellites in orbit looking down from space for explosions on Earth. As a side benefit, they see asteroids entering the atmosphere. We now know that a hundred tons of asteroids and meteors hit the earth every day, almost all of them small and harmlessly vaporized.

By 2007, enough cross-disciplinary analysis had been performed to conclude that a massive asteroid had definitely hit the Yucatán Peninsula near Chicxulub at the end of the Cretaceous Period. In fact, scientists have come to define the end of the Cretaceous Period itself as the moment the Yucatán asteroid hit the Earth, the Cretaceous-Paleogene extinction event. The explosive impact almost defies measurement. Scientists talk of a detonation nearly three billion times bigger than the Hiroshima and Nagasaki atomic bombs combined. The result was earthquakes, tsunamis, fires, acid rain, darkness, and enough dust and debris thrown up into the atmosphere to completely alter the Earth's climate for decades or even hundreds of years. Three quarters of the life-forms on the planet went extinct, according to most estimates. Morrison noted in *Cosmic Catastrophes* that this extinction event is only one of several in the fossil and geological record. He wondered how many of them had been caused by asteroid impacts.

With his dry humor, Morrison suggested there was actually a silver lining to the Cretaceous-Paleogene extinction event. It killed off all the dinosaurs that couldn't fly, thus making possible the rise of mammals, including humans. Otherwise, he noted, sixty million years later it might have been a dinosaur descendant, some sort of intelligent lizard, sitting in a space agency in North America trying to get other lizards to care about the possibility of an asteroid impact. Without Chicxulub as evidence, it would be even harder for the lizard to mount his case.

The story of the asteroid killing the dinosaurs and almost all other forms of life is chilling at one level, but upon reflection the probability of experiencing such an event seems tremendously remote. It did happen well before humans even existed. It seems as unlikely to be a problem for us as dinosaurs themselves are. The evidence that Morrison and others offered was of events that, in the lifetime of an average human, seem highly improbable. His account of a contemporary killer meteor hitting Russia was only fiction, designed to help the reader visualize what could happen, because there was no real-world event in our lifetimes that would inform us. And then, in 2013, there was. Once again we return to Russia.

It was a Friday morning in February, cold but clear. In Chelyabinsk, a city of a million people beyond the Ural Mountains on the edge of Siberia, most people were at work or school by the time it happened at 9:20 a.m. The flash of light streaking across the sky was brighter than the sun, so bright, in fact, that some people later claimed it gave them a sunburn that caused their skin to flake off. The noise, the heavy clap of a sonic boom, came next. A wall of pressure broke windows and collapsed roofs. Flying glass injured hundreds.

If, as Morrison had suggested twenty-five years earlier, the asteroid exploding above Russia had happened at the height of the Cold War when the Soviet Union's nuclear missiles were on hair-trigger alert, people might well have assumed it was an American nuclear attack. In fact, the Chelyabinsk explosion of 2013 is now estimated to have been twenty times greater than the American nuclear attack on Hiroshima. Chelyabinsk, however, was not leveled. The damage was minimal because the explosion took place thirty kilometers above and to the west of the city, as the sixty-five-foot, twelve-thousand-ton asteroid hit the atmosphere and vaporized. Only some football-size meteorites survived, scattered over hundreds of square kilometers.

Astronomers had been expecting a near miss (maybe more appropriately called an "almost hit") that day from the asteroid they had been tracking. They expected it to pass just outside the Earth's atmosphere and continue off into space. And it did, sixteen hours after the other asteroid that they didn't see exploded over Chelyabinsk. The big rock that sent hundreds of Russians to the emergency rooms had snuck up on Earth, coming out of a part of the sky where the astronomers were blinded by the Sun. When it became clear that there had been an asteroid "sneak attack," it grabbed the attention not only of astronomers, but government leaders in Russia, the United States, the United Nations, and elsewhere. It proved that we do not know enough about the location and trajectories of all local asteroids to be confident that we can always predict an impact.

Surprises are common in space. On the evening of October 25, 2016, the Panoramic Survey Telescope and Rapid Response System (Pan-STARRS) on Maui found an object of between 5 and 25 meters in diameter heading toward Earth. They calculated that it would hit our planet, or *just* miss it, in five days. Obviously it missed, but it was yet another reminder that many objects are not seen until they are very close. Paul Chodas of NASA's Jet Propulsion Laboratory estimates that, as of late 2016, only about 25 to 30 percent of asteroids in the 140-meter class had been found. At that size, an impact would destroy a city.

The first time humans had ever seen an object coming and predicted an impact had been only eight years earlier. On October 6, 2008, an astronomer in Arizona spotted an object on an Earth impact trajectory and alerted other observers. They calculated that the incoming rock would hit within a day, likely in northern Sudan in nineteen hours. They were precisely accurate as to both time and place. The impact area was unpopulated. The rock was small. There was no damage. Many scientists were pleased at this predictive ability, which had never been demonstrated before. Others, however, focused on the relatively few hours of advance warning that they had been able to give, less than one day.

In 1988, Morrison and his colleagues were unsure of the exact probability of a large asteroid striking the Earth, but they did their best to estimate it. Drawing on an unpublished report of a 1981 NASA workshop at Snowmass, Colorado, Morrison suggested that the probability of a civilization-extinction-level impact (like Chicxulub) was one in three hundred thousand a year. As unlikely as that still makes it sound, he wrote, for each individual it is ten times more likely than dying from a tornado, and we do have tornado warning systems. Nor should we examine only the extinction scenario. The likelihood of an impact causing major global effects short of civilization extinction is much higher. Even more likely is an asteroid impact that wipes out an entire metropolitan area. There was, however, an admitted imprecision in the estimates, because at the time there was inadequate data regarding the number and size of asteroids and meteoroids that might become near-Earth objects.

Morrison made a plea for the systems needed to get better data, telescopes and radars dedicated to scanning space for asteroids and meteoroids that are potentially threatening to human life. He wrote:

> *We have used some pretty dry words and numbers to relate*
> *probabilities about the most awful imaginable catastrophe,*
> *the very destruction of civilization as we know it. Thousands*
> *of years of history erased in a single day . . . Should we shrug*
> *our shoulders and watch the next TV program? Or should we*
> *rally the nations of the world to mount a Project Spacewatch to*
> *remove this threat hanging over us? [an] effort . . . protecting*
> *civilization against the risks of a probably preventable accident?*

Morrison was part of a group that rallied nations to act, particularly the United States. Between Morrison's book in 1988 and the impact over Chelyabinsk in 2013, a lot had happened. Astronomers around the world, NASA managers, the media, moviemakers, and even Congressmen had taken his warnings seriously.

In 1990, two years after *Cosmic Catastrophes* was published, Morrison testified before Congress about the threat of an asteroid impact. In the House of Representatives Science Committee he had a sympathetic audience. The committee directed NASA to study the issue. Led by Congressman George Brown of California, over the following years the committee repeatedly urged NASA to further investigate the asteroid threat.

THE SPACEGUARD GOAL

In 1992, Morrison penned NASA's answer to the committee request. In his Spaceguard Survey Report, Morrison suggested that NASA have a goal of discovering 90 percent of the near-Earth asteroids a kilometer or more in size during the next twenty-five years, by 2017. In 2016, he explained to us that he didn't believe it was possible to discover 100 percent of near-Earth asteroids, thus the goal of only 90 percent. Later, as Spaceguard revised its goals and looked for smaller asteroids, NASA maintained a 90 percent metric as the target. It is realistic and scientifically based, but the 90 percent detection metric is an admission that there will always be a threat of a surprise.

In calling the report Spaceguard, Morrison was tipping his hat to the great science-fiction author Arthur C. Clarke. In his novel *Rendezvous with Rama*, published in 1973, Clarke demonstrated his deep understanding of astrophysics and engineering, as he had decades earlier when he became the first to suggest the concept of an orbital communications-relay satellite. In *Rama*, Clarke had written that Earth had established the Spaceguard surveillance system after an asteroid impact had devastated northeastern Italy in the year 2077.

Thus, it was Morrison who suggested to Congress that they task NASA to do the study, and it was Morrison who then conducted the research and suggested the "Spaceguard Goal," which, in 1994, the House then directed NASA to achieve. Four years later, NASA announced that it would heed this House request. The space agency may have felt the need to do so not because of its employee David Morrison or its Congressional overseer George Brown, but because of an additional influence.

Hollywood had gotten into the act. In 1998, not one, but two blockbuster movies hit the big screen with the premise that the Earth was about to be hit by a giant object from space. The first release was *Deep Impact*, which had the benefit of a NASA scientist acting as a consultant. Not surprisingly, David Morrison thinks it was the better of the two films and "very realistic." (One perhaps unrealistic aspect for 1998 was that the U.S. President was an African American.) In the film, an eleven-kilometer comet is on a collision course with

Earth. The U.S. and Russia team up, sending astronauts to intercept it and destroy it with nuclear weapons. When the bomb explodes, the comet splits into two pieces, one of which hits off the coast of North Carolina, causing a massive tsunami that floods U.S. coastal cities. The astronauts heroically sacrifice themselves to destroy the second, larger piece before it can slam into Earth and cause an extinction-level event.

In July came *Armageddon*, in which asteroids began hitting Earth (specifically New York City), causing astronomers to detect a "Texas-size" asteroid only eighteen days away from Earth impact. Before that happens, Paris and Shanghai are hit by precursor strikes (apparently these asteroids are city seekers). NASA responds by launching two shuttles with blue-collar wildcat oil-drilling teams to land on the big rock and drill holes into which they will place nuclear bombs. In the end, of course, after many mishaps, the bombs go off and save the planet.

The two films had amazing sales around the world, producing almost a billion dollars in box office revenue. They made their studios rich. They also caused people everywhere to wonder how realistic an asteroid threat really is. NASA's survey work began in earnest, led by astronomers at MIT's Lincoln Lab, based near Boston but using a telescope in New Mexico. Their work, and that of the Mount Lemmon SkyCenter in Tucson and other centers worldwide, fed into the Minor Planet Center that NASA had funded in Cambridge, Massachusetts. They were able to estimate that the number of asteroids a kilometer or larger in near-Earth orbit probably was about a thousand.

Not satisfied, Morrison pushed for a new Spaceguard goal. He wanted to look for smaller asteroids, those between 140 meters and one kilometer (460 to 3,200 feet) in diameter. Such intermediate-size rocks could still be devastating, especially in the highly improbable case that one directly hit a city. Six years after their book *Cosmic Catastrophes*, Morrison and Chapman published a scientific paper that argued that our concern should be not just with the civilization-extinction impacts that might occur once in a million years, but also with smaller impacts that could create a major regional catastrophe, such as wiping out a major city. NASA estimated that there could be as many as thirteen thousand near-Earth-orbit asteroids in that size class. Just one would cause an explosion equivalent to 129 megatons (a thousand Hiroshima-size detonations) and dig out a crater over two kilometers across. Therefore, by 2003 NASA had declared an additional goal of detecting 90 percent of asteroids over 140 meters.

Again, it took Congress two years to direct NASA to conduct such a survey. It did so in the George Brown Act, named after the former House Science Committee chair, and signed by President George W. Bush. Now the question for NASA was how to execute the research, while the question for Congress was how much it was willing to pay. NASA and the National Science Foundation joined with other nations and organizations to jointly propose the Large Synoptic Survey Telescope (LSST). It is now under construction in Chile and is set to begin full operations in 2020. Were LSST to be used exclusively for the near-Earth asteroid mission, it alone could map 90 percent of near-Earth asteroids larger than 140 meters by 2037. Meanwhile, though, the U.S. Air Force is building a complementary system, Pan-STARRS, in cooperation with Lincoln Lab and the University of Hawaii. The first of four telescopes is operational, but the completion date of the entire system is unclear due to funding difficulties. A large portion of both LSST and Pan-STARRS's time will be used looking for near-Earth asteroids.

Other scientists have been pushing, so far unsuccessfully, for a space-based telescope dedicated to planetary defense. In 2013, they got an unplanned space-based system, at least briefly. From 2009 to 2011, the Wide-field Infrared Survey Explorer (WISE) satellite mapped the stars. However, WISE eventually ran out of the coolant that it required to look deep into space. It was placed in hibernation. After the previously mentioned Chelyabinsk impact in Russia, NASA began looking for ways to step up its search for asteroids. There was a realization that the WISE satellite, then sleeping, could contribute to the effort without the need for coolant. Thus, in August 2013, the satellite was turned back on and dubbed NEOWISE. It has since identified over forty potentially hazardous asteroids, those in near-Earth orbit and large enough to be of concern.

Using all of these systems at least part time, NASA finds about eight near-Earth asteroids a month. Knowing the location of only a handful of these objects in 1988 when Morrison suggested they start looking, NASA and other organizations have since found over 14,000 of them. Among that number are more than 880 with a diameter greater than a kilometer, objects that could potentially wipe out human civilization. The NASA consensus is that there are approximately 70 more that have yet to be located. The initial Spaceguard goal of finding 90 percent of these larger asteroids has been achieved. Morrison says it has all been "very successful." He has described the work to date as "a very good job of finding the biggest asteroids over the last fifteen years" and, as a result, we now know that "we are much safer than we thought we were" when the surveying started. None of the larger asteroids that have been located seems to be headed for Earth anytime soon. David Morrison now says that we can be fairly confident that those extinction-size rocks will not hit us, at least not in the next century or so.

Although civilization is not at risk from a large asteroid in our lifetime, some cities may be. Morrison notes that the Tunguska asteroid of 1908 was ninety meters in diameter. If a similar-size object exploded over San Francisco, he suggests, it would kill tens of thousands. If it hit a few hundred miles east in the Nevada Desert, "it would be a tourist attraction." There are thought to be over a million near-Earth asteroids of about that size. Well over 90 percent of them have not been surveyed. We do not know where they are, nor if and when they will strike Earth.

And then there are the comets. Unlike the asteroids, whose orbits are well known, long-period comets, those with an orbital cycle of two hundred years or longer, are difficult to predict. While some asteroids' paths can be projected for decades, even centuries, comets can still surprise us, suddenly showing up within Jupiter's orbit, only months away from Earth. Traveling at three times the speed of an asteroid, a comet of the same mass would strike a planet with nine times the energy of the slower-moving space objects.

In 2016, NASA announced the creation of a Planetary Defense Program Office to coordinate efforts to find and react to significant-size asteroids and comets on possible Earth-impact trajectories. Also in 2016, NASA announced its intention to launch an Asteroid Redirect Mission (ARM) as soon as 2020. The concept is, well, creative. It envisions an autonomous spaceship traveling to an asteroid, landing on it, finding a large boulder, attaching the boulder to the spaceship, launching from the asteroid with the boulder, and then flying alongside the asteroid for months, using the gravity of the spaceship/boulder combination to gradually change the asteroid's path.

Morrison, however, calls the ARM project "strange" and told us he "has no high hopes for it" being fully funded or ever working. He points out that it was not a mission originally designed to protect Earth from dangerous asteroids, but rather to explore an asteroid with the thought of perhaps mining it for rare metals. It was repurposed as a defense mission to gain support for its funding. The fact that people in NASA now think sticking a Planetary Defense label on a program will help get it funded speaks volumes about how far we have come from the early days when Morrison was begging NASA to start looking for asteroids.

PASSING THROUGH THE NEXT FILTER

Thus, we might think that David Morrison's efforts have been successful. Indeed, he told us that he and his colleagues have "accomplished a lot more than I ever thought we would." Those don't sound like the words of a man who considers himself a Cassandra, one still struggling to get the world to take his issue seriously before it's too late. "People did call me Dr. Doom," Morrison says, initially because of his pioneering work on the asteroid threat. The name has stuck, but it is now used because he is the astrophysicist most likely to respond to the many "ridiculous" scare stories about other threats from space. He spends a lot of his time debunking myths and dispelling fears about remote future dangers, such as the Sun becoming a red giant and engulfing the earth. ("It will," he says, but not for many millions of years, and humans will have either vanished as a species or moved to other planets by then anyway.)

We think David Morrison's Cassandra Coefficient is moderately high, even though he may not. The issue seems like the stuff of science fiction or Hollywood, in part because it has been. While the Earth has been hit before by large objects that did catastrophic damage, it was very long ago and these events are exceedingly rare. So long ago was the last extinction event that it is fair to think that the warnings now face Initial Occurrence Syndrome, because such an event has never truly happened in modern human history.

The audience, however, has been responsive, up to a point. Over thirty years, Morrison notes, there has been significant progress in gaining acceptance by the scientific community. No one really disagrees with him on probabilities or consequences. Even governments have acted. At least, as he observed, one government has acted by slightly opening its treasury, enough to map many of the possible asteroid threats. No government, however, has acted to give Earth a comprehensive defense system that can rapidly spring into action to deflect a large and threatening object, were one to be found.

Morrison himself exhibits many of the traits of a Cassandra. He is a renowned expert, uses the scientific method, and is data driven. He was the first to see something and say something about it, loudly. Although a government employee, he influenced the system to get Congress to tell his agency what to do. He has not experienced the criticism of his colleagues as overtly as others. No one doubts that he's right, but some have other priorities. People don't want their projects derailed, especially by preparing for an unlikely catastrophe.

Morrison differs from other Cassandras in that he is relatively accepting of the satisficing solutions that the system has offered up to deal with his concerns. He has not moved on from the issue, but he has found time to make a lot of progress in other areas of astrophysics and space exploration. His obsession with planetary defense is well under control, perhaps because others have taken up the cause and seem more focused on the need for a deflection or defense capability.

His theses gained widespread peer acceptance relatively quickly, for which he says he was "gratified." But, he adds, it *did* take six years for NASA to accept his first proposed Spaceguard goal for surveying the asteroids near earth. It took decades to conduct the initial survey and will take more decades to do the follow-up surveys to find the smaller but still potentially deadly asteroids. Nonetheless, his warnings were sufficiently heeded that the U.S. government spent hundreds of millions of dollars on that asteroid survey work. There is, however, still much to survey in the class of asteroids that could destroy a city. Almost thirty-five years after Morrison's seminal warning in *Cosmic Catastrophes*, there are hundreds of thousands of near-Earth asteroids that could wipe out a city, but are still not located.

Like those of many other Cassandras, Morrison's warning has not really overcome institutional reluctance. Because he has always had at least one Congressional backer, he has been able to force NASA to address the asteroid issue, but not to the point of creating a preventive capability. If tomorrow we were blinded by the sun to a 140-meter asteroid coming at us, one that hit with little or no warning and wiped out London, would we think that the governments of Earth had responded appropriately to David Morrison's warning of 1988? Or would we see Morrison as a Cassandra, someone who had seen a disaster coming, who was an expert in his field, who was data driven, but still was paid only lip service, was given low priority, and was ultimately unable to stop a true cosmic catastrophe?

While discussing the "strange" Asteroid Redirect Mission proposal, Dick asked Morrison if he was bothered by the fact that there is no capability to deflect an incoming asteroid. "Yes, of course," he said. "It's kind of silly to wait to build that capability until you see an asteroid coming."

If he didn't think much of the ARM concept of lifting a boulder off an asteroid, what would he propose? "Kinetic impactor," he responded quickly, an Earth-launched missile that would intercept an asteroid and change its trajectory by the force of the collision and perhaps by an explosion creating a shock wave. Would that work? "In principle yes, but we could not really be sure" that a specific system would do the job without testing it. Developing it and testing it could "take at least ten years," and we have not yet started.

Why not? Morrison cited three reasons, the first two of which are recurring problems for Cassandras. First, NASA has a much smaller budget than it once did (using a constant-dollar comparison), and it has other priorities for its limited funds. This is one form of the phenomenon we in this book are calling institutional reluctance. It also illustrates the "Pet Rock problem," not wanting to defund an organization's chosen projects, or Pet Rocks, in order to deal with some unpleasant, unplanned, and unwelcome obligation. Organizations faced with this kind of intrusion into their priorities usually think that somebody else should fund efforts at risk reduction, that it should come from some "national budget," not theirs. It's a classic example of the bystander effect, or a diffusion of responsibility.

Second, Morrison said, we are talking about "a hypothetical danger." There has never been an asteroid impact with any significant number of casualties in human history. This is a clear example of what we in this book are calling Initial Occurrence Syndrome, the failure to value a risk sufficiently even in the face of convincing data, because "it has never happened before."

Third, Morrison adds, there is a political threshold problem. It might not be until an asteroid is about to strike Earth that we could predict with high confidence where the impact would take place. If we don't know that there is a high probability of an asteroid hitting a U.S. city, or a city we value elsewhere (like London), will there be enough political interest to act? If it's likely to hit the ocean, a desert, or in one of our least favorite countries, why should we bother to do the extraordinary? The problem, Morrison says, is compounded because "you are always going to be dealing with uncertainty." You probably won't be able to call the White House and say that a particular asteroid is on a trajectory to hit an American city. Even if you could, by the time you were confident of the location of impact, it would probably be too late to develop the kinetic impactor or anything else to provide planetary defense.

Calling the White House is a problem for him even now. Morrison is not allowed by NASA to pick up the phone and call the Executive Office of the President to lobby for funding. Only NASA

Headquarters can do that, and even then only in support of the budget priorities in NASA's official budget request (a budget request that does not include research and development of a kinetic impactor or any real system of planetary defense). Morrison admitted to me, "I have never had a conversation with anyone in the White House like the one that you and I are having now" about the remaining risks and the need for a response capability.

To David Morrison's three reasons why nothing is being done to fund planetary defense, we might add a fourth. The asteroid-threat response so far is a clear-cut example of institutional satisficing. Sixty years ago, when Nobel Laureate economist Herbert Simon coined the term *satisficing*, he thought it usually occurred when decision makers lacked complete data. Thirty years later, he added that "the peripheral vision of a complex organization is limited, so that responses to novelty in the environment may be made in inappropriate and quasi-automatic ways that cause major failure."

In the case of planetary defense from near-Earth asteroids, NASA may be satisficing. The agency, in response to Congressional pressure, has funded an asteroid survey, although it has taken decades to make significant progress, is not complete, and never received the total allocation of any telescope. Data on the risk of an Earth impact is uncertain. Nonetheless, NASA can say it is acting. It now has an office named Planetary Defense. It has repurposed an asteroid exploration and mining mission for planetary defense. What is has not done is plan, design, request funding for, test, or deploy any system to respond to an Earth-threatening asteroid. In response to "novelty in the environment" then, NASA may be responding in a "way that could cause major failure."

If the United States doesn't build a planetary defense capability, no one else will. Despite all the meetings of UN committees on the near-Earth asteroid threat, despite the concerns of astronomers in the European Union and elsewhere, despite the wake-up call Russia had at Chelyabinsk, "no other country has acted," according to Morrison. The U.S. has found every asteroid found to date. The U.S. manages the system that tracks them. Other nations talk a lot about the issue, but "they haven't found one damn asteroid" so far, and many lack the technological expertise and resources to do so. No nation other than the United States can be expected to develop a capability to intercept an Earth-threatening asteroid, but for now at least, it won't either.

In the meantime, Morrison believes, "Nature is more likely to draw attention to the issue" with another surprise visitor. He hopes that, when the next one comes, it won't wipe out a city or kick up so much dust as to block sunlight and cause a nuclear winter effect. Despite his frustration, Morrison has not given up. He believes that the key to catalyzing more progress on this issue is to portray the asteroid risk more clearly. Perhaps NASA Headquarters or the Executive Office of the President will come to terms with the magnitude of the threat and fund a planetary defense capability. With his NASA colleagues, he wants to develop "a better way to present the risk, a better hazard analysis" methodology. Why? Because "it all comes down to, How much risk are you willing to take?"

The risk we are running now is beginning to be recognized more broadly. In October 2016, the U.S. government's disaster planning and response unit (the Federal Emergency Management Agency) held a simulation exercise of an asteroid crisis. Teamed with that agency were representatives of NASA and the state of California. The scenario was simple, a 100 meter long (330 foot) asteroid had been discovered headed for Earth with a predicted impact in 2020 on Los Angeles. In the exercise, NASA stipulated that four years was not enough time to design, build, launch, and utilize a deflector device that would push the asteroid aside. Therefore, in the game the decision was made to start evacuating Los Angeles immediately, in part because a complete relocation of ten million people and all of the functions they perform might take four years.

The major take-away from that exercise should have been that evacuation was necessary because NASA has not seriously started on a defense capability. The risk Morrison has warned of has not sufficiently penetrated the minds of the decision makers to result in funding a response capability.[3]

David Morrison and the Asteroid Threat

CASSANDRA CHARACTERISTICS	THE WARNING	THE DECISION MAKERS	THE CASSANDRA	THE CRITICS
High			X	
Moderate	X	X		X
Low				

CHAPTER 16

The Biologist: Gene Editing

But Natural Selection, as we shall hereafter see,
is a power incessantly ready for action, and is as
immeasurably superior to man's feeble efforts, as the
works of Nature are to those of Art.

—CHARLES DARWIN, *ON THE ORIGIN OF SPECIES*

The column of soldiers emerged from the chilly night. Each face similarly hardened and bare from the long day of training and the many preceding. Each face exactly the same. Identical. These American infantrymen were designed in a lab, their biology optimized for perception and endurance, and deployed to the field in 2088. The same was true for pilots, submariners, intelligence specialists. Governments from Beijing to Moscow to Washington viewed genetic optimization as critical to national security; and individuals from Paris to Seoul to New York viewed it as critical to social and financial security. Wealthy parents could not only specify their child's eye and hair color but splice in genes to improve athletic and academic prowess. They owed their thanks to North Korea.

The kleptocracy in Pyongyang had long since collapsed, but in the first half of the twenty-first century North Korean scientists surreptitiously perfected a low-cost and efficient technique to modify the genes of human embryos. Known as CRISPR/Cas9, it was previously used ubiquitously and admirably around the world, curing humankind's genetic diseases, helping livestock grow faster and larger, and making crops more nutritious. International agreements prohibited any human enhancement not deemed medically necessary. But North Korea was never one to play by the rules, and in 2036 it began secretly selling the technology to global elites that wanted to give their children an unmatched advantage: superior DNA.

By that time, the world had become accustomed to using CRISPR/ Cas9 to fix human disease. So when word leaked out about genetically enhanced CRISPR babies in the early 2040s, few were truly surprised, and the international regulatory infrastructure quickly collapsed. CRISPR clinics, prohibitively expensive for the vast majority, rapidly became the most profitable sector of the global medical industry. Governments adopted CRISPR as a matter of national security. A new class of elite human would come to dominate the species. This was never the intent of the woman who had created the technology in 2012, but the thought had crossed her mind, and haunted her nightmares.

Dr. Jennifer Doudna was raised in Hilo, Hawaii, a small, rainy, seaside town on the biggest—but one of the most sparsely populated— islands in the chain. Doudna moved there with her family after her father took a job as an English professor at the University of Hawaii. Bookish from an early age and surrounded by tropical plants and animals, she developed an intense fascination with how the world works. What were the processes responsible for the abundant life around her? From rain-forest mushrooms to tide-pool seashells, that fascination continues to drive her career to this day.

"My dad loved books, and he loved reading about science," she told us when we sat down in her office at UC Berkeley, where she is a professor of biology and manages a huge laboratory. "He would buy books and throw them on my bed. Early on, I was probably in about sixth grade or so, one of them was *The Double Helix*. I remember reading it and being just stunned that scientists could devise experiments to figure out something like the structure of DNA. It was just mind-blowing to me as a kid."[1]

From then on, it wasn't enough to just observe nature. "I wanted to know the *chemicals* that were in there; what's the difference in the DNA that made, for example, one mushroom different from another?" Professor Doudna laughed, recalling the memory, and it was clear that the spark of scientific wonder, first realized in those early years, remains just as strong today. "Coupled with the knowledge that you can do experiments to understand those chemical differences, I thought, *Yes, I want to spend my life doing that.*"

Beside a striking view of San Francisco Bay, the walls and shelves of her office are filled with awards, accolades, and memorabilia that would serve as the enviable record of any successful scientist's career. But the fifty-one-year-old is nowhere near retirement, her career a relentless crescendo of scientific discovery. In 2016, she's one of the world's most famous scientists. Many believe her discovery of CRISPR will send her to Stockholm to win a Nobel Prize.

Professor Doudna's early interest in science became an academic passion, which was guided by mentors and peers toward a deeper interest in biochemistry. First, at Pomona College, she worked in the lab of Sharon Panasenko, her undergraduate advisor, whom Doudna credits as a strong female role model in science. She later completed her Ph.D. at Harvard under Jack Szostak, who went on to share the 2009 Nobel Prize for his work in genetics. There Doudna developed a fascination for RNA, the close relative to its more famous genetic cousin, DNA.

DNA, as is taught in any high school biology class, is the molecule that carries the hereditary information of all living organisms. Its famous double-helix structure looks like a long spiral staircase, twisting around itself, the steps holding the encoded information in what are called base pairs. The order of those base pairs is what's important; it's a blueprint for our design, regulating the creation and functions of all our cells and making life as we know it possible.

While Doudna was at Harvard, RNA was widely believed to be DNA's secretary: RNA carried out the commands spelled out by DNA. But there was a growing belief that RNA did more than just run DNA's errands and fetch its coffee, and Doudna's doctoral and postdoctoral work focused on this new theory. Soon, it became clear that RNA actually was responsible for various functions, and some types of RNA could even act as enzymes, catalyzing reactions.[2]

Working at the University of Colorado as a postdoctoral researcher, Doudna crystallized these "ribozymes" (RNA molecules that act as enzymes). Just like Watson and Crick did before her with DNA, Doudna used X-ray diffraction to reveal the ribozyme's three-dimensional structure. "When I was in grad school, I remember thinking, I would not want to do that," Professor Doudna said, with a laugh. But scientists increasingly believed that this catalytic RNA folds itself into complex, three-dimensional shapes and may have been involved in the origin of life. "It was a fascinating question, but at that time, nobody had ever seen what any of these RNA molecules looked like. I suddenly realized that I had to do crystallography, so I could actually see them."

It was an unprecedented breakthrough in the field. In 2000, at the age of thirty-five, she received the prestigious Alan T. Waterman Award from the National Science Foundation. Established by Congress, the Waterman Award is given to a single researcher each year for an outstanding contribution to science or engineering. By then,

Professor Doudna had taken a position at Yale, where she strengthened her reputation as a focused, creative, and eminently brilliant researcher with keen attention to detail. In 2002, she became a professor at the University of California at Berkeley, continuing her work by investigating the RNA of viruses.[3]

EVOLUTION, BY UNNATURAL SELECTION

Although Professor Doudna was focusing on viruses, in 2005 a colleague at Berkeley, Dr. Jill Banfield, approached her regarding some work she had been doing with unique bacteria found in a highly acidic abandoned mine. The genetic material of many of these bacteria, Dr. Banfield said, contains numerous "clustered, regularly interspaced, short, palindromic repeats," or CRISPR for short. In essence, some of the DNA repeats itself. Over the years, scientists realized that there are "spacers" between these repeated sections, and that the spacer consist of sequences of DNA from viruses that had previously invaded the bacteria. Scientists discovered that the genetic code stored between these repeats plays a special role in some sort of bacterial immune system, allowing the cell's defense mechanisms to quickly identify a re-invading virus and cut up its genetic material.

Scientists were getting tantalizingly close to a breakthrough but hadn't yet solved the key puzzle that would revolutionize genetic engineering. Then, in 2012, the revolution began. Professor Doudna had been approached by Dr. Emmanuelle Charpentier at a microbiology conference the year before. Dr. Charpentier, a French scientist working at Umea University in Sweden, was studying the genome of flesh-eating bacteria. She and her colleagues had been investigating the process by which bacterial CRISPR sequences, coupled to a CRISPR-associated (Cas) protein, protect the cell from viral invasion.

Charpentier hoped to recruit the well-known structural microbiologist to their effort and tease out the structure of the CRISPR/Cas complex. Dr. Doudna agreed.

After about a year's work, Doudna and Charpentier's teams had determined that in a specific type of CRISPR/Cas system, type-II, two strands of RNA copied from the bacteria's genetic material combine with a protein referred to as Cas9. One strand of RNA acts as a guide, matching and targeting invading viral DNA; the other is an activating strand that doesn't itself bind to the target DNA but is necessary for the complex to function properly. Altogether, this CRISPR/Cas9 structure efficiently and accurately slices up the matching DNA from an invading virus.[4]

The eureka moment—when, as Professor Doudna describes it, "the project went from being 'This is cool, this is wonky' to 'Whoa, this could be transformative'"—came when the scientists discovered that the complex was still effective even if they artificially engineered those two strands of RNA as a single strand using relatively simple techniques.[5] Put simply, Professor Doudna, Dr. Charpentier, and their teams had created a tool that could slice open the DNA of any organism at any point they specified, and the artificial CRISPR/Cas9 complex worked just as efficiently and effectively as it did in nature. At that moment, they realized that their exploration had transcended basic research and had become a genetic tool with the power to reshape and engineer any known DNA-based life-form.

Scientists have known for decades how to paste together different strands of DNA once they've already been cut using an enzyme called DNA ligase. What was previously missing was a method of easily, accurately, and cheaply cutting existing strands of DNA. CRISPR/Cas9 is that method, and it can be used successfully to edit the DNA of any type of cell, including plant and animal.

In one way or another, humans have genetically modified organisms for their benefit for millennia. The earliest agrarians cultivated crops and raised animals, selecting for seed and breeding stock those with more desirable traits, such as plants that produced larger, better tasting, and more nutritious fruits or vegetables. Selectively breeding these crops over time resulted in genetic evolution that favored those traits. That's how hearty kernels of modern corn arose from the seeds of wild plants, how wild Southeast Asian birds became domesticated egg-laying factories, and how wolves evolved into man's best friend, not to mention the designer Labradoodle or Cockapoo canine mixes in vogue today. In all of these instances, favorable traits encoded in the underlying genetic material were combined with other favorable traits and passed on to future generations, resulting in more and more desirable organisms.[6]

The revolutionary aspect of CRISPR/Cas9 is that scientists can now potentially isolate and edit genes in a single generation that would have traditionally taken scores of generations to change. Scientists can splice genes into crops that impart drought, pest, and disease resistance. They can modify genes in livestock embryos so they'll grow into adults remarkably faster, be more resistant to pathogens, and provide healthier meat. Already, Chinese scientists have used the technique in goat embryos to delete genes that suppress hair and muscle growth, resulting in animals with longer hair and more muscle, ostensibly better at producing both meat and wool. In fact, dozens of Chinese labs have plunged headlong into CRISPR/Cas9 experimentation in fields ranging from animals to agriculture to biomedicine to human transformation.[7]

Personalized medicine and the elimination of genetic disorders is another potential bright spot in the future of CRISPR/Cas9. Scientists can now snip out defective genes that cause disease, replacing them with healthy ones. Such a procedure would be most easily carried out for conditions caused by a single genetic mutation, such as sickle-cell anemia, Huntington's disease, or cystic fibrosis. A CRISPR-based therapy would substitute a properly functioning gene for the defective section of DNA in the patient's cells.

In an extraordinary 2013 experiment, scientists at the Massachusetts Institute of Technology used this technique to treat mice suffering from tyrosinemia, a rare liver disorder that also affects humans. A single mutation in the genetic code prevents the liver from producing an important enzyme, which causes toxins to build up in the body and often requires a liver transplant. Using CRISPR/Cas9, the researchers were able to splice functioning DNA sequences into the mouse liver cells, restoring proper liver function.[8] Similarly, researchers at the University of Texas Southwestern Medical Center successfully used CRISPR/Cas9 in 2015 to treat Duchenne muscular dystrophy, a more common incurable disease that affects about 1 in every 3,500 boys and causes the body's muscles to break down, resulting in paralysis and premature death. In experiments using mice, the scientists removed defective sequences of DNA that interrupted cells' production of an essential muscle protein, partly alleviating the condition, and giving hope that a cure, while years away, could one day be possible.[9]

While fixing human disorders caused by multiple genes—like amyotrophic lateral sclerosis (ALS, or Lou Gehrig's disease) and many types of cancer—presents additional complications, CRISPR/Cas9 is already being used by medical researchers to engineer away analogous diseases in animals. These animal models are facilitating more accurate and effective research, helping to identify exactly which genes are involved and how, and accelerating identification of therapies that might be most effective at treating the comparable disease in humans. Beyond that, the hope remains that eventually CRISPR could result in a permanent cure.[10]

The possibilities are exciting and potentially worth billions of dollars. Biotech companies devoted to developing CRISPR-based therapies have already launched around the world, backed by millions in venture capital. Caribou Biosciences, founded in Berkeley by Professor Doudna, and CRISPR Therapeutics, a Basel-based startup founded by Dr. Charpentier, are two of the earliest commercial ventures in the nascent field.

However, while CRISPR may be the key to developing revolutionary solutions to many of the world's maladies, it may also prove a dangerous and divisive force shaping the future of humanity.

FROM GERMS TO THE GERM LINE

What might happen if viruses, modified to cause aggressive cancer in lab animals for medical research, accidentally get out of the lab and start a cancer pandemic? What happens to the food chain if a genetic modification to prevent malaria also causes malaria-carrying mosquitos to become infertile, causing a worldwide mosquito population collapse? And considering the public outcry about classically genetically modified organisms, how freaked out will (and should) the public be when scientists start using CRISPR to tweak the DNA of our grocery-store produce, making salmon a brighter orange and giving tomatoes and cucumbers a longer shelf life?[11]

Another serious concern arises from what are known as off-target events. After its discovery, researchers found that the CRISPR/Cas9 complex sometimes bonds to and cuts the target DNA at unintended locations. Particularly when dealing with human cells, they found that sometimes as many as five nucleotides were mismatched between the guide and target DNA. What might the consequences be if a DNA segment is improperly cut and put back together? What sorts of effects could this cause, both immediately and further down the road for heritable traits? Experimenting with plants or mouse bacteria in a controlled laboratory environment is one thing, but what is the acceptable level of error if and when researchers begin experimenting with a tool that cuts up a person's DNA? If an error is in fact made, is there any potential way to fix the mistake?[12]

There are also important considerations about the types of cells that are genetically modified. Genetic modification could one day be used to treat disease in an organism's somatic, or body, cells. But those somatic genetic edits would remain only with that single individual, passing away with its death, and would not affect future generations. Using CRISPR to treat a woman suffering from tyrosinemia, for example, would genetically modify her liver cells, but her children, should she later become pregnant, would lack the improved gene. However, CRISPR can also modify the DNA of a fertilized egg or embryo. As its cells divide and mature into a fully-formed human being, every cell, including those that might one day be used for reproduction, will contain a copy of that modified DNA. Such "germ-line" modification means that any changes would be passed on to that individual's offspring, becoming a part of the broader human genome.

Few would likely object if such a modification could prevent a child from developing a horrible disease like muscular dystrophy, but what if the treatment introduces unexpected complications, perhaps from an off-target event, that don't manifest until later in life, or maybe even a generation later? To prevent their modified DNA from mixing into the broader human genome, can an individual be prohibited from procreating? Would such a restriction even be feasible, let alone ethical?

These questions aren't simply thought experiments. In 2014, researchers in Nanjing, China, used CRISPR/Cas9 to carry out gene editing in the embryos of cynomolgus monkeys, the first time the technique had been used successfully in primates. Cynomolgus monkeys are frequently used in biomedical research to model human disease because of their genetic similarity to people. The experiment's overall success rate was low, but it nonetheless foreshadows the direction of future experimentation: primates now, and human experimentation next.[13]

While her research focused on some of the smallest molecules present in microscopic bacteria and viruses, Professor Doudna knew the implications of CRISPR were huge. She began growing uneasy with each new revelation pointing toward the inevitable ramifications of the tool she had helped to create. At first she believed the questions were better left to others. "I'm not a human geneticist," she stressed, her career devoted to basic science research. Leave it to others to hash out the nuances of how to regulate CRISPR's use, particularly in humans. But reticence is not a quality that's built into Professor Doudna's DNA.

"I would say it was the monkeys," Professor Doudna responded when we asked if there was a tipping point that compelled her to sound a warning about CRISPR/Cas9. "In February of 2014 a reporter called me here, in this office, to ask me about [the cynomolgus monkey] experiment. I thought, *Wow, this is moving so fast!* I had a very strong feeling of the disconnect between what the public knew and what all of us that were involved in this technology were doing." Despite not having a technical background in bioethics or public policy, she felt an obligation to act.

"I even thought, probably our government, our regulators, don't have any idea that this is going on. And yet, maybe one of these weeks I'm going to get a phone call that someone's just made a CRISPR baby. And that thought really freaked me out." Professor Doudna paused, then continued, "I said to myself, it's not responsible to just sit back and say that's the purview of others. Who better than those of us engaged in the science to explain what CRISPR is, what it does, and where it's going." She, along with fifteen of her colleagues, organized a conference to discuss the technology. At the forefront of the agenda was what to do about what Professor Doudna feared was the rapidly approaching next step: "It was becoming clear that this technology worked in any kind of cell. The logical next step was, well, can you use it in a human embryo?"

THE MOTHER OF CRISPR BECOMES ITS CASSANDRA

As the group of biologists gathered in Napa, California, in late January 2015, several couldn't help but recognize the similarities to a conference that had taken place almost exactly forty years earlier.

In February 1975, about 150 leading professionals gathered at the Asilomar Conference Grounds that overlooks the Pacific Ocean on California's Monterey Peninsula. The meeting had been called to discuss a recent breakthrough discovery that allowed scientists to artificially manipulate the genome. Those in attendance were mostly molecular biologists, but the broad implications and wide-ranging discussions also brought physicians, lawyers, journalists, and government policy makers to Asilomar.[14]

The topic of discussion was recombinant DNA technology. Several years earlier, scientists had discovered restriction enzymes, enzymes that cut DNA at a single, specific sequence of nucleotides. Each restriction enzyme is specific to a certain DNA sequence, whereas CRISPR can be modified and customized to cut DNA wherever you like. Around the same time, scientists had also discovered an enzyme called DNA ligase, which pastes strands of DNA together. Biologists now had the basic tools to slice up two different strands of DNA and put them together.[15]

In 1972, Stanford biologist Dr. Paul Berg and his student, Janet Mertz, did exactly that, combining two separate molecules of DNA into a single strand, creating the first artificially engineered recombinant DNA molecule. But concern over safety issues gave Berg pause. Scientists debated the biohazard risks that would arise if an engineered organism, like a highly virulent, cancer-causing bacterium, escaped from the laboratory. Some questioned if recombinant DNA experiments should be carried out at all.[16]

Three years later, Dr. Berg was the chairman of the Organizing Committee for the International Conference on Recombinant DNA, charged with developing guidelines and recommendations as humans forged ahead with genetic experimentation. Helping Berg organize the conference was the renowned MIT biologist Dr. David Baltimore, himself a powerful force in the study of DNA and RNA.

The conference produced a set of recommendations and voluntary guidelines encapsulated in a summary statement that was delivered to the National Institutes of Health and published for the rest of the scientific community in the Proceedings of the National Academy of Sciences. It classified experiments based on risk and delineated an escalating set of precautionary measures that should be taken as the risks increased. The statement also called for a voluntary worldwide moratorium on genetic experimentation that could severely endanger public health, such as those involving highly pathogenic organisms and toxic genes.[17]

The guidelines served as the basis for the National Institutes of Health guidelines that later governed use of the technology, and they have functioned as a successful template for the safe undertaking of recombinant DNA experiments ever since. Fears that recombinant DNA experiments could unleash a public health disaster were never realized. Moreover, early excitement at the promise of recombinant

DNA also gave way to the reality that manipulating DNA, precisely specifying the cutting location, proved surprisingly tricky. It remained that way until Professor Doudna's CRISPR breakthrough.

Still, Asilomar is credited with serving an even more important role. Dr. Berg explained to us in his Stanford office, where he still serves as a professor emeritus, that "what Asilomar accomplished was establishing trust between the public and the science." Over 10 percent of the attendees were from the media, "who were there as participants," he stressed, "not just as observers." The journalists took part in all of the discussions, asked questions of the panelists, joined in for late-night beer drinking and debating with the scientists and bioethicists, and were given the freedom to write about the conference as they saw fit. The articles were detailed and dense, including a feature in *Rolling Stone*, recounting the debates, the arguing, bickering, and the ultimate consensus. The result, Dr. Berg recalled, was that "the public concern just fell off."[18]

Many credit these successes to the Asilomar Conference's timing: standards could be set and broad consensus could be reached because scientists involved in the technology's conception called for an early, comprehensive, and public discussion. Had they waited longer, they would have been trying to forge new rules amid myriad different procedures among labs throughout the world. In addition, involving the public helped to head off any misinformed pseudoscientific backlash.

Forty years later, Berg and Baltimore must have been struck by the irony of their situation. While effective in ensuring the safety of recombinant DNA technology experimentation for the years to come, the revolution many had predicted early on was stymied by the lack of a more efficient way to cut DNA. Now here they were, that tool had been discovered, and the scientific community again needed to figure out a way forward. Like Dr. Berg and recombinant DNA technology, Professor Doudna, the inventor of CRISPR, was

now a leader in the effort to understand and prevent the possible unintended consequences that could result from its unfettered deployment and adoption.

Given the type of experimentation already underway using
CRISPR, the questions the scientists at Napa tackled were markedly
different from those at Asilomar. "We never discussed ethics," Dr. Berg
told us, "and we did it on purpose." The darker questions were still
beyond the horizon, and biohazard concerns were paramount at the
time. While Asilomar focused on establishing broad safety protocols,
those gathered at Napa discussed the risks of modifying the human
genome.

Professor Doudna and the others in attendance saw their Napa
conference as a prelude to a broader international and public dialogue
on the practical, ethical, social, and legal implications of CRISPR.
The two-page commentary they published called for a worldwide
moratorium on any use of CRISPR for human germ-line editing. But
to inform more robust consideration of CRISPR's potential and risks,
they also encouraged the continuation and expansion of ongoing basic research that might one day be applicable to engineering human
genetics.

Just as significantly, the Napa group called for better communication and discussion about CRISPR. Doudna, Berg, Baltimore,
and the other attendees believed that an open and frank discussion
among the public, scientists, legal experts, bioethicists, and other
stakeholders was necessary to tackle these issues. Their commentary
concluded by recalling Asilomar, "at the dawn of the recombinant
DNA era, the most important lesson learned was that public trust in
science ultimately begins with and requires ongoing transparency and
open discussion. That lesson is amplified today with the emergence
of CRISPR-Cas9 technology and the imminent prospects for genome
engineering. Initiating these fascinating and challenging discussions

now will optimize the decisions society will make at the advent of a new era in biology and genetics."[19]

Though the meeting was small, its participants were influential. The Napa attendees hoped that their commentary would buy some time in forging a broader consensus on the use of CRISPR. At the very least, they hoped human germ-line editing remained simply a fiction.

Two weeks after the Napa commentary was published, Chinese scientists at Sun Yat-sen University reported that they had used CRISPR/Cas9 to artificially alter the DNA of human embryos in an attempt to correct a gene that was responsible for a rare blood disorder. The Chinese study used embryos that were tripronuclear, containing three sets of DNA, and could not develop into a human being in an attempt to sidestep the ethical questions of editing the human germ line. The study was released in an obscure, Chinese-published journal called *Protein & Cell* after the mainstream publications for such research, *Nature* and *Science*, rejected the paper, considering the research unethical.[20]

The experiment wasn't a great success. Of the eighty-six embryos they started with, the Chinese researchers found that CRISPR/Cas9 made the correct edits in only four of them. All four were found to have off-target events, suggesting that the treatment would not have worked if they had been trying to edit viable human embryos. But the fact that scientists had genuinely made an alteration of the DNA of human embryos at all was itself significant.[21] If nothing else, the study did exactly what Professor Doudna hoped to avoid: it thrust CRISPR into the public consciousness in an unsettling and alarming way. The *Washington Post* carried the headline "The Rumors Were True: Scientists Edited the Genomes of Human Embryos for the First Time"; the *New York Times*: "Chinese Scientists Edit Genes of Human Embryos, Raising Concerns."

Professor Doudna and her colleagues upped their public outreach campaign, granting magazine and newspaper interview requests, publishing commentaries, even testifying to Congress and officials in the White House. Her efforts succeeded in attracting the attention of legislators. "I sensed a desire to both understand the technology and to understand if this was a technology that needed new regulation," Professor Doudna said of her briefing to the House Committee on Science and Technology.

On her second trip to Congress, Professor Doudna and some of her colleagues involved with CRISPR held an all-day public meeting in Washington, D.C., attended by members of Congress, staffers, students, and the general public. "The astounding thing there was that it was standing-room only for the entire day," she recounted. "I had a sense that people were realizing we are on the verge of being able to edit the human genome in a way that will change evolution, that something profound was happening in science."

Their main message remained that CRISPR should not yet be used to modify the human germ line and that, while research on treating human diseases should certainly continue, the public should be aware that there are risks and benefits of such therapy and that many answers to the relevant questions remained unknown. Policy makers were engaged. The public was interested. Perhaps Professor Doudna will avoid Cassandra's fate?

A CRISPR FUTURE

Despite the failure of the Chinese scientists' experiment with human embryos, other researchers are already making significant strides in reducing CRISPR's error rate. Some believe that improvements to the technology will soon render it safe enough for experimentation on humans, if we haven't reached that point already. Dr. George

Church, a geneticist at Harvard and a signatory of the commentary that laid out the Napa conference's recommendations, estimates that advances could yield an error rate of about 1 in 300 trillion base-pairs (for reference, the entire human genome is 1/100,000 that number), about the same rate that genetic errors occur naturally.

Still, darker questions surrounding the future applications of CRISPR remain at the margin, and not just in the minds of conspiracy theorists. In a November 2015 interview with the *New Yorker*, Professor Doudna said she recently dreamed that a colleague of hers took her to meet someone. "I went into a room and there was Hitler. He had a pig face and I could only see him from behind and he was taking notes and he said, 'I want to understand the uses and implications of this amazing technology.' I woke up in a cold sweat. And that dream has haunted me from that day. Because suppose somebody like Hitler had access to this—we can only imagine the kind of horrible uses he could put it to."[22]

CRISPR might be the key to one day curing formerly terminal or life-altering diseases, but what about other traits: Asperger's, a tendency to overeat, or even an introverted personality? We are entering an era where CRISPR will be used to splice genes into embryos to improve IQ, select for taller children, and enhance athletic prowess. CRISPR will very likely, one day soon, give parents the option of choosing the color of their children's eyes, skin, and hair. The sequencing of the human genome coupled with the power of CRISPR/Cas9 has brought eugenics forward from an uncomfortable fiction. Even traits that are influenced by hundreds or thousands of genes will, in the not too distant future, be decoded.

Once researchers identify these genes, where does society draw the line between conditions for which gene editing is and isn't appropriate? And will it even remain possible to maintain prohibitions on experimentation?

Consider these two interrelated questions: First, should it be legal to genetically modify embryos to select for IQ, height, eye color, and other improvements? Second, if it were legal and safe, would you elect to use these enhancements to improve your own child?

At the Napa meeting, Professor Doudna recalled someone saying, "'There may come a time when, ethically, we can't *not* do this.' That kind of made everybody sit back and think about it differently." Imagine a future when CRISPR is error-proof. If a doctor were to discover that an embryo would be born with a life-threatening disease, would he or she then feel compelled to utilize CRISPR to correct the relevant portions of the embryo's DNA?[23]

In 2003, after thirteen years of international scientific collaboration, the entire three-billion-base-pair human genome was sequenced, at a cost of nearly $3 billion. Today the technology has improved so much that sequencing an individual's genome costs about $1,000 and can be done in about a day.

Dr. Steve Hsu of Michigan State University believes that these advances will allow us to identify the genes responsible for intelligence within ten years. While intelligence is probably related to some ten thousand different parts of the genetic code, Dr. Hsu believes that we can begin to find correlations by comparing different individuals' genes and measures of intelligence, like standardized testing scores. Aggregate enough of this data and a statistical analysis will permit you to say, with a high level of certainty, which loci of genetic variation correlate with superior intelligence. While not perfect, this aggregation of data will allow scientists to find genetic correlations for any number of traits that parents might want to optimize in their offspring: athleticism, visual acuity, or reduced likelihood of developing Alzheimer's.[24]

Such are the social concerns to which Professor Doudna and the Napa group alluded. Simple in vitro fertilization currently costs upwards of $10,000, but gene editing would likely cost many times more. Could CRISPR become a tool for the wealthy elites to engineer genetically superior babies? In a world already racked by increasing wealth inequality and an increasingly distinct cultural divide between the rich and poor, will genetic enhancement accelerate and make permanent the already yawning gap between the haves and the have nots? In the 1997 movie *Gattaca*, genetic information and modification became a tool used to correct imperfections while stratifying the masses into de facto castes. Could CRISPR irreversibly and forever create a privileged segment of society with superior intelligence, looks, and health via a self-perpetuating cycle of self-selection?

Previous advances in genetic technology also raised the specter of eugenics, but never before has the tool been as simple, efficient, and promising as CRISPR/Cas9. If the elites can harness the power of gene editing for the benefit of their offspring, governments too could harness the capability as a means to bolster the country's power, maximizing its human capital, and more specifically, its intellectual capacity. Failing to exploit CRISPR could become a national security liability.

Jamie Metzl is a novelist and biotechnology policy expert at the Atlantic Council. Formerly a colleague of ours on the White House National Security Council, Metzl foresees a day when human genetic enhancement becomes an integral, if not the most important, part of the international arms race. He has written two novels exploring the realpolitik imperatives behind controlling and exploiting the technology that would impart an unmatched competitive edge. Human beings genetically designed to be better scientists and engineers, mathematicians and data analysts, policy makers, even superior soldiers.

Already the precedent has been set, albeit on a much smaller scale. "China already identifies promising athletes as young children, takes them from their parents, and raises them to be Olympians,"

Metzl told us over lunch. "This is the next evolution." More ominously, "once the Chinese start doing it, what's to prevent others from letting the ethical considerations go out the window in order to keep up?"[25] From a game-theory perspective, it is easy to imagine how all countries will have an incentive to engage in the practice if it were to provide some measure of a societal advantage.

An engineered army of supersoldiers controlled by elite decision makers with genetically enhanced logic and problem-solving skills is still decades away, if it ever becomes a reality, but Professor Doudna and many of her colleagues still fear that the wider public is unprepared to understand, let alone debate, all of the implications of CRISPR/Cas9. These questions aren't relevant only to Americans, the conversation must transcend national boundaries.

In December 2015, ten months after the Napa Conference and nearly eight months after the Chinese scientists published their paper on editing human embryos with CRISPR, the International Summit on Human Gene Editing was held in Washington, D.C., sponsored by the U.S. National Academy of Sciences and National Academy of Medicine, the Chinese Academy of Science, and the Royal Academy of the United Kingdom. As the Napa Conference was meant to be only a starting point for a further discussion on the future and implications of CRISPR/Cas9, several members of the Napa Conference, including Drs. Doudna, Baltimore, and Berg, sat on the summit's organizing committee.

Over three days, scientists, policy makers, bioethicists, lawyers, and journalists discussed somatic-cell and germ-line modification, international governance of CRISPR technology, societal implications of genetic editing technology, and next steps. Would CRISPR result in a more equal and healthy society, or would it exacerbate global inequality? Are current international bodies and mechanisms capable of establishing a global governance infrastructure for the oversight and use of CRISPR? Or is international oversight simply a pipe dream?[26]

Like the Napa Conference, the summit produced opposing viewpoints and few clear answers. And like the Napa Conference, the summit released a final statement to serve as a set of interim guidelines for another future, more robust round of ethical and regulatory debate. It called for ongoing basic research and prudent precautions in moving toward any therapeutic use of CRISPR/Cas9 in human somatic cells, as well as for an ongoing international discussion on the role and use of gene editing technology in the future.[27] Professor Doudna believes it's a good first step. "The meeting was incredibly helpful for people like me in the United States to understand cultural differences with my colleagues in other countries," she said. And as officials around the world begin to debate if and how CRISPR should be regulated, scientists too have to continue along a parallel track, "trying to agree on how we're going to use the science."

Her efforts are also motivated by concern that public backlash resulting from a shocking CRISPR experiment would have a chilling effect on the entire field, cutting funding and resources across the board. "I want to avoid something getting published, maybe regarding human embryos, that really attracts negative public attention and results in regulation that would inhibit innovation and legitimate research," she said. "I think that scientists can go a long way towards avoiding that negative outcome by at least engaging, being a part of a public discussion, even if we don't have all the answers."

"The Pandora's box is open," she acknowledged. "The thing about this technology that is so wonderful but also concerning is that it's so easy to use. And even if regulators said today that we couldn't use CRISPR anymore, how would they ever enforce that? So we're hoping to invite scientists in to be a part of the conversation," to create norms of behavior rather than regulation, and to keep the nonscientific world engaged in the dialogue.

The momentum of her recent efforts has helped blunt some concerns, but she hopes policy makers, regulators, scientists, and others continue to "keep a very active eye on this technology." Simply relying on hope isn't an effective strategy. "Science is global now. People are clever and creative. And now all sorts of things are possible."

If CRISPR keeps getting better and easier, how can we guarantee that someone won't go rogue and do some kind of previously unthinkable human experiment? Professor Doudna paused and turned to look out at San Francisco Bay, glittering under the sun of an unusually clear day. "The thing that is holding us back is our understanding of the human genome. And that will change over time, of course." She still fears that one day she will wake up to news that a CRISPR baby has been born. But she hopes that the efforts of herself and others are beginning to set the stage for society to understand and discuss the risks and benefits of CRISPR and to plan accordingly.

TOWARD A BRAVE NEW WORLD

The unintended potential consequences of CRISPR and genetic engineering have a moderate Cassandra coefficient. While the issue is complex and has aspects of being considered outlandish, researchers and bioethicists have already initiated robust discussions with policy makers on the implications of the technology. The headline-grabbing nature of CRISPR has already begun to generate significant public awareness, which these same scientists recognize must be informed by open and accurate information from academia.

Professor Doudna's willingness and ability to constructively engage with others, an attribute many of the other Cassandras in this book lack, is a reflection of her equally important job as an educator, in addition to researcher. Her ease with communication and ability to explain even tough scientific concepts embodied what she didn't even need to tell us: "I like people!" She also attributes it, in part, to her upbringing in a diverse, multicultural town in Hawaii, something she misses since she has left the islands. "I learned to appreciate that people from different cultures view life differently, and we should learn about it, understand it, and talk about it with one another. And that's how I've approached this whole scientific conversation: 'I know this, you know that, let's get together, and let's have a conversation.'"

While our Cassandra is a proven and respected scientist and is largely responsible for the creation of the technology, she freely admits, "I'm somewhat of an outsider to the genomics community. My research has always been in fundamental biology and biochemistry." Unlike other Cassandras, who lacked the social power to get their message across, Professor Doudna told me, "I was far enough in my career that I knew a lot of people already and they knew me"—which is not a small understatement. Her connections helped her quickly rally others to the cause. Those who joined in her concerns were some of the world's most respected biologists and bioethicists.

Concern arises from the unknowns: whether and how CRISPR could be used to edit human cells, if it will be used to edit the germ line, if it will be used to enhance rather than simply fix. An unknown also arises from how effectively a new set of global norms around CRISPR will be adopted by the international community. As we have seen with artificial intelligence, pandemic disease, and other warnings, diffusion of responsibility is a real challenge. Can comprehensive oversight of CRISPR even be exercised? Like so many other emerging scenarios, there exists no one clear set of decision makers. Oversight and control of CRISPR's use must come from a consortium of actors: companies, universities, governments, and international bodies. How might those regulations change when CRISPR proves to be a game changer, because of its lucrative potential or because of its initiation of a new global arms race?

It no longer seems to be a question of *if* a CRISPR baby will be born but instead *when*. When it does happen, what kind of brave new world will that baby be born into?

Dr. Doudna and Gene Editing

CASSANDRA CHARACTERISTICS	THE WARNING	THE DECISION MAKERS	THE CASSANDRA	THE CRITICS
High	X	X	X	
Moderate				X
Low				

CHAPTER 17

Can You Hear Her Now?

I n the preceding seven chapters, we have applied our Cassandra Coefficient, analyzing a confluence of factors to determine if certain experts are likely to be Cassandras, if their warnings are both correct and unheeded. In each case, we have identified the pertinent characteristics of the issue, the audience of decision makers, the Cassandra, and the critics. The responses to these warnings range from the promising, like the concerted effort to create global norms around the use of CRISPR/Cas9, to the piddling, like the nascent struggle against artificial intelligence's possible threat to humankind.

Our intent was not to take sides, but rather to say that these possible Cassandras and their warnings should be heard. More than just being given a pro forma audience, they and others like them should have their predictions adjudicated in a thoughtful and comprehensive system. To do so is not just a favor to Cassandra. It's a potential contribution to the safety and security of society at large.

OTHER POSSIBLE CURRENT CASSANDRAS

This book is not an exhaustive study of every warning with the potential for a high Cassandra Coefficient. For example, former House Speaker Newt Gingrich and former CIA Director James Woolsey both publicly worry about the threat from an electromagnetic pulse (EMP), the burst of energy released by a nuclear explosion. In addition to its powerful blast and radioactive fallout, nuclear weapons can also damage or destroy electric wiring, from power lines to computer circuits. If detonated high in the atmosphere, the damage from an EMP would be widespread. A Congressionally mandated commission investigated the EMP issue and issued stark warnings in 2008.

Mitigation steps, like coating wires and shielding data centers, would reduce or prevent the EMP effect, but they are expensive. Thus far, no U.S. administration has been sufficiently moved to budget the resources necessary for significant mitigation or prevention. Although neither Gingrich nor Woolsey are proven technical experts or First Order Thinkers about this, as Cassandras tend to be, they are carrying the message of those that are. They claim an EMP attack would be devastating. Others believe the risk is exaggerated. The debate continues.

Another warning questions the United States' preparedness for a catastrophic earthquake in the Pacific Northwest. Chris Goldfinger, a professor of geology at Oregon State University, has studied this topic extensively and concluded that the area is long overdue for a cataclysmic earthquake. He predicts a one-in-three chance the quake will strike within the next fifty years. Seven million people reside in the anticipated devastation zone. Imagine the Boeing factory near Seattle, the biggest building on Earth, brought to the ground. Imagine the pools of spent nuclear fuel in the Hanford Nuclear Reservation contaminating groundwater with radioactive waste for thousands of years. And Goldfinger says this earthquake is not a question of if, but when.

The estimated magnitude of the destruction is difficult to appreciate and the mitigation costs immense, partly explaining why we have not yet acted. In fact, we are not prepared in the slightest. The last major earthquake in the Pacific Northwest occurred around 1701, well before written records in the area. Scientists began to understand the magnitude of the looming disaster only in the last couple of decades, but still, despite empirical evidence, safeguards and emergency response remain almost nonexistent. Earthquake-proof structures would need to become the norm. Plans for quickly restoring electricity would need to be written. Expensive assets would need to be relocated. Nuclear and ballistic missile facilities would need to be hardened. The clock is ticking.

DOOM AND GLOOM, OR MAKING PROGRESS POSSIBLE

Some readers will, by now, no doubt accuse us of being obsessed with the unlikely, purveyors of doom and gloom. Actually, we are still both very optimistic people with great faith in science and engineering, confidence in the potential of leaders to effect change, and hope that the future will be better than the past. Nonetheless, that brighter, better future will not occur by itself. Progress is not inevitable.

A desire to ensure progress leads us to the critical recommendation of this book: institutionalizing systems to deal with Cassandras, within universities, corporations, governments, and international bodies. Such systems must expressly encompass four functions. First, they must scan the horizon, alert for new warnings that might otherwise go undetected. Second, they must sift the credible from the dubious, separating the signal from the noise. Third, the systems should employ a consistent methodology to evaluate possible courses of action in response to the warning. And fourth, they must include a strategy that effectively implements the desired response.

As senior advisors to people like U.S. President Bill Clinton and UN Secretary General Kofi Annan, we were faced with identifying and developing responses to issues such as terrorism, weapons proliferation, cybersecurity, and the global HIV epidemic. While many of these problems were never fully or, in our view, successfully addressed, our shared experiences have given us some ideas about how we might structure a system to better deal with disasters in the offing. Using those lessons and the cases studies presented in this book, we suggest addressing the four necessary functions as follows:

1. SCANNING FOR PROBLEMS

Reliably listening for and identifying warnings requires an institutional solution. In our experience, the typical decision maker in any organization already has a demanding job that leaves little time to regularly conduct broad searches for additional problems. We believe an office whose specific mission is future forecasting and warning is the most effective way to carry out this function. The solution we discuss here is for the U.S. government, but other institutions could adopt something similar.

Some parts of the government already have processes for identifying future problems, but warning as a function is not embraced nor institutionalized. The U.S. Public Health Service has the little known but highly important Epidemic Intelligence Service, whose staff are popularly known as "disease detectives" and look for new and emerging diseases. Most of the time, however, a government department or agency sets up an office to address a problem only after someone discovers it. NASA established a space surveillance program following David Morrison's repeated warnings about asteroids. Charlie Allen's office was set up specifically to centralize within the bureaucracy an entity that could provide a warning of war only after the attack on Pearl Harbor. It was disbanded decades later.

Trouble arises when no part of the government is looking for potential and unforeseen problems, when an office that should be paying attention is already task-saturated with its other responsibilities, or blinded by conventional wisdom from seeing initial occurrences or the unusual. Worse yet is when the relevant agency puts an urgent problem in the queue for routine consideration in due course. In the United States, Congressional oversight committees are theoretically responsible for rectifying such trouble. In practice, however, they typically spend little time doing actual oversight, unless they see the potential to embarrass the other political party.

The White House, with its decision-making power, is best suited for creating and hosting such a forecasting office. We are unaware of any point in the past when such a function has been formalized. Some with warning messages have managed to get through to White House staff through persistent effort or, usually, personal connections. Eliezer Yudkowsky is not the first to raise the specter of artificial intelligence. Almost twenty years ago, the computer scientist Bill Joy used his influence in Silicon Valley to get a meeting with the White House Chief of Staff to raise concerns over artificial intelligence and robotics. Despite his meeting, nothing happened. His warning seemed to be about something that was outlandish, or was at best well over the horizon and could be dealt with at some future date, if ever.

A formalized National Warning Office, in the Executive Office of the President, would be the administration's focal point for identifying disasters on the horizon. This small, elite team should not be, as Charlie Allen's office was, part of the intelligence community, although it could task intelligence agencies to collect and analyze information. Rather, the office would have a broad, even intentionally vague, mandate to look across all departmental boundaries for new and emerging threats. The office should not address ongoing, chronic problems, such as obesity. Rather, the focus should be on possible impending disasters that are not being addressed by any part of government.

The National Warning Office should also work through the interagency and the White House on two institutional goals: first, to create management and decision-making environments that nurture, rather than stamp out, dissent and warning from qualified experts; and second, to develop a small cadre of people drawn from every cabinet agency to establish processes and information sharing to recognize sentinel intelligence.

Fortune 500 companies, and industry sectors, especially those in the critical infrastructure, could create futures groups to scan for largely unnoticed risks, particularly new ones.

On an international level, the UN Secretary General might also attempt to create such a team, though our experience with the UN suggests that no new office there remains small or elite for long.

2. SEPARATING THE SIGNAL FROM THE NOISE

How would such an office identify problems worthy of consideration and differentiate them from the many unworthy? The work we have done for this book convinces us that the process should, in part, be to listen to the people giving warnings. Professional experts giving warnings do not hide under rocks. They publish in peer-reviewed journals and deliver papers at academic conferences. If they are constrained by working in a government agency, they will, if they become as obsessed as our Cassandras, contact the national office we propose.

Unlike those who favor statistical analysis or the development of people into being good predictors, and quite unlike those who think that we simply have to accept that "stuff happens," we think that we can discriminate among the warnings by first focusing on the people giving them and applying our Cassandra Coefficient. Do they have the characteristics of our Cassandras? Are they proven experts in the field? Do they have professionally developed data? Do they lack a record of past, incorrect warning? Is there an absence of any personal monetary or other gain, e.g., working for an industry that would benefit?

Comparing the issue and the audience against the characteristics of the Cassandra Coefficient is the next step. If the individual, issue, and audience together give the problem a high Cassandra Coefficient, the national office would place the issue on the agenda for "fast-failure" review. Such a review would include examining the data and bringing in a team of relevant experts to dialogue with the Cassandra and his or her critics in an attempt to comprehensively assess the validity of the prediction and examine any faults or inconsistencies in it.

Should the issue hold up under that intense scrutiny, the national office might then provide the results of that review to a standing senior advisory panel, a multidisciplinary part-time group, which would review the fast-failure evaluation, make additional inquiries or request further information, and advise whether or not the issue should be taken up for immediate action by a senior interagency decision-making group. In an extreme case, the fast-failure review could result in the issuance of an immediate formal warning distributed to the President and his senior staff.

One example of disaster prediction that needed objective analysis involves President George W. Bush and the Iraq War. Bush says he was told by people he thought were experts that Iraq was developing weapons of mass destruction that would be used against the United States or Israel. His advisors said that the cost of a war to prevent such weapons from being built would be minimal and would pale in comparison to the cost of waiting until after a WMD attack.

We believe that using our Cassandra Coefficient would have shown a rational decision maker that those providing the warning were not true Cassandras and that their data was wrong. They were not experts in the Middle East. Indeed, the regional experts were appalled at what Bush's advisors were proposing. A national office conducting a true review of the "evidence" would have found that much of it was fabricated or exaggerated, as some intelligence professionals were saying at the time.

This monumentally ill-informed decision aside, if a Cassandra warning is judged valid, then what? As the case study of Charlie Allen and the invasion of Kuwait demonstrated, even having a Warning Staff and someone willing to issue a Warning of War may not be enough to move decision makers. A system for formulating and evaluating response strategies is also needed, one informed by some familiarity with types of responses that may not have worked well in the past.

3. RESPONSES: SURVEILLANCE, HEDGING, MITIGATING, AND PREVENTING

The third function in our system involves generating and rigorously evaluating potential responses to the Cassandra's problem and establishing a ballpark estimate of the resources it will require. Any executive, in the White House or elsewhere, who receives a warning will immediately ask for the recommended course of action, or better yet the set of options, to deal with the disaster.

We believe the best way to accomplish this would be a coordinated effort, led by the national office in concert with whatever agencies are implicated in the issue. The national office's key contribution and responsibility should be to bring standardized intellectual rigor in formulating and evaluating the different responses using devil's advocates and opposition teams to examine the problem and proposed solutions from multiple perspectives. The implicated agencies bring with them subject-matter expertise and the ability to reach out to their networks of experts in relevant fields. Together, these teams would formulate response options to deal with Cassandra's warning, interconnected strategies that fall along a spectrum of options: surveillance, hedging, mitigation, or prevention.

Before diving in, a word of caution. In chapter 9 we identified satisficing as one of the characteristics that may be associated with a Cassandra Event. Because many of the problems we examined have costly mitigation or prevention strategies, maybe even requiring substantial changes in the way we live, doing just enough to quiet demands for action, while failing to actually solve the problem, tends to be the easiest of solutions. By definition, satisficing solutions are insufficient but attractive responses. They must be avoided through the candid analysis we describe below.

* **SURVEILLANCE:** The first category of response strategies, and the one that requires the least amount of effort to actively counter the possible disaster, is surveillance. A surveillance strategy is similar to the intelligence community's I&W (indications and warnings) process that we discussed in chapter 2, monitoring and measuring tells that correspond to increasing levels of risk or likelihood that the disaster will occur. If a warning proves to be of real concern but more pieces would need to fall into place before it could be

considered a threat deserving a response, the national office should issue collection priorities to the relevant cabinet agencies to watch and report back on those specific missing pieces. To warn in advance of what would otherwise be a surprise jump in sea level, for example, the national office perhaps would direct measurement of key indicators, such as melting from the bottom of the glacier in the Amundsen Sea portion of the West Antarctic ice complex. The office would work with private-sector and nonlinear thinkers to seek to add creative indicators to potential threats deemed worthy of being on the office's watching brief.

Such a response is sufficient only if scenario modeling and other analytical techniques have shown that, when indications of a possible risk reach a certain level, there will be time to react. This judgment becomes crucial. How much time will we have? Professor Jennifer Doudna hopes that current efforts to establish norms around the use of CRISPR/Cas9 technology will succeed. Perhaps monitoring the use of the technology will be enough. In the case of a large Earth-bound asteroid, David Morrison warns us that there may be little or no time to react. James Hansen fears that the time to act to avert disastrous global sea-level rise is almost past.

While engaging in increased surveillance and measurement, the national office and its partnered agencies should also begin analyses of alternative responses and their critical paths. Determining responses, figuring out how much they might cost, and developing an understanding of when they might need to be implemented are equally important parts of the surveillance strategy.

* **HEDGING:** These actions flow immediately into the next
   category of responses to a potential disaster: hedging. It
   coexists with surveillance, but is specifically focused on
   investing resources into getting ready for more robust
   mitigation or prevention responses. It's an interim phase that
   consists of ongoing monitoring with preparation until the
   surveillance system determines that active measures must
   begin.

   In addition to knowing when to pull the trigger, in
   a hedging strategy the key question becomes, as Alain
   Enthoven asked in the title of his groundbreaking book on
   defense budgeting in 1971, *How Much Is Enough?* When
   determining how much is enough, governments turn to
   analysts who do cost-effectiveness studies, usually placing a
   monetary value on human lives. In making these kinds of
   resource decisions, it is also useful to inject a little real-world
   thinking regarding politics, the media, and public reaction.

   David Morrison was troubled by the cost-benefit analysis
   that judged the risk from an asteroid strike as the number of
   total deaths from a single incident averaged over the number
   of years between strikes. Such a technique suggested few
   deaths per year and, thus, the risk did not warrant significant
   expenditure. In the real world, no one will take comfort
   in such a calculation following a major asteroid strike that
   kills millions. Countering Islamist terrorism, if looked
   at from a cost-benefit accounting perspective, is greatly
   over-resourced. Since the September 11th attacks, very
   few people have died in the United States from Islamist
   extremist terrorism. Right-wing domestic groups have killed
   about twice as many people in this country over the same

period. This sort of number crunching misses the political reality that the way people are killed, and the concomitant way we are forced to live, matters.

Thus, some kinds of possible disastrous events should be given the needed resources as a matter of public policy, despite coldly calculated metrics. For in corporations, issues that draw public scrutiny or criticism may not, in fact, greatly affect the bottom line, yet leaders who can't show they had done everything possible to stop a disaster are frequently treated harshly by the electorate, by history, or by the shareholders.

One way in which to establish the appropriate level of expenditure is to begin by determining what it might cost to deal with the threat, and *then* compare that to what it would cost to implement a rebuilding or restoration program if the disaster were to occur. Hedging strategies involve determining how much funding should be allocated to prepare for a disaster while laying the groundwork for future steps to reduce the effects or prevent it entirely.

* **MITIGATION AND PREVENTION:** Prevention of a disaster is the best course of action when feasible in time and cost. Mitigation is the rational response if prevention is not possible or guaranteed. However, even when pursuing a prevention strategy, mitigation should still be considered as additional assurance. These costs are typically small compared to the cost of rebuilding after a disaster occurs, but in the absence of reliable probability data, making a decision entirely on quantitative grounds can be difficult.

Knowing when the data is rich and extensive enough to trust it and when it is too scant is difficult. If data is in short supply, don't worry about probability. Probability is not likely to be a useful measure when dealing with the risk of a disaster that has never occurred before, or only happened at great intervals. Even when there are regular occurrences, it's important to test whether new factors have been introduced that change the probability. Climate change may now be making the "hundred-year flood" take place every few years.

Instead, focus on possibility. Is it possible? Could it happen? What would have to occur to make it happen? What is there to stop it? How confident are we that those preventative systems would work? How well have we tested them? Good analysis must also imagine how a prevention measure could fail or be circumvented. In the Ukrainian power-grid failure that agitated Joe Weiss, there were visual and aural alarms in the control room. The hackers took over the control panels to show that everything was normal when, in fact, the system was collapsing.

In Cassandra Events, the common yardstick of "reasonableness" when calculating risk may be unreasonable. Worst-case scenarios are often excluded from cost calculations because they never happened before or are statistically unlikely. However, if the worst case is not physically impossible, then its actual costs should still be taken into account. There is a growing realization among risk managers that the probability of many kinds of significant events is essentially unknowable. How often will an asteroid strike Earth, a terrorist group fly aircraft into skyscrapers, or a new virus strain create an epidemic? The data are insufficient for such predictions, but worst-case scenarios do happen.

Failure to account for the costs of the worst-case scenario

is a common refrain from the first half of this book. In every case presented, the costs of rebuilding and restoration after the disaster far exceeded what mitigation or prevention efforts would have cost, often by orders of magnitude. Such "economies of prevention" range from tangible and monetary, like higher sea walls around Fukushima Daiichi or a more robust ventilation plan at Upper Big Branch, to personal and political, such as understanding the damning math of Markopolos's Madoff investigation or arming the Syrian opposition. In retrospect, the choice is obvious. Our hope is that the system we have outlined here can help to clarify these choices in the future before the disaster strikes, before a warning becomes a Cassandra Event.

4. PERSUADE

The fourth and final part of the system we propose focuses on successfully recommending and implementing the response strategy, one that might be controversial, expensive, disruptive, or perhaps all three. Just as Cassandra found it impossible to warn the people of Troy, the national office will likely face a heavy burden to persuade the leadership to take a (previously ignored) warning seriously.

Warnings may always meet with flat-out rejection or satisficing. The national office will need to ensure the creation of an effective strategy to build consensus and commitment among the decision makers, appropriators, the public, and other nations. A strong persuasion campaign will be rich in inarguable data and should recruit credible supporters along the way. Ideally, the relevant cabinet agencies will take over this persuasion role as soon as possible.

The primary determinate for successfully carrying out the response is the quality of the leader. Our combined six decades of experience in the government and private sector tells us that only leaders who are convinced of the urgency of an issue get things done. Whether president of a country or a company, the executive must ultimately have confidence that he or she is taking the best course of action after thoroughly assessing all of the available evidence.

Sir Winston Churchill, as we noted in chapter 1, was a Cassandra. His dire warnings couldn't prevent World War II, but when the disaster came, his ability to inspire and lead his people was the sine qua non of victory. He didn't minimize the threat when speaking to his people. He didn't sugarcoat the sacrifice that would be required. He offered to die in the pursuit of the course of action he proposed. After his warnings were proven true, the people found no reason to doubt his assessment or his determination. Never has a leader been more demanding or persuasive when proposing how to deal with a disaster.

In the absence of a Churchill, we have seen over the years a few leadership techniques that have been used with varying degrees of success in times of crisis. Surprising people, presenting both a novel problem and the bill to fix it, is a likely recipe for rejection. Allowing (or working with) experts, the media, and legislators to expose the problem, and allowing the public to demand solutions, is frequently a better choice than letting the government be first to identify the problem and simultaneously propose the solution.

While implementation of the strategy requires decisive direction from the executive, the national office should have a continuing responsibility for providing oversight, periodic evaluation, and updates of the implementation efforts. The office should take into account changes in the situation since the response decision was made in order to ensure that the strategy remains effective over time.

VULNERABILITIES BURIED BY COMPLEXITY

In addition to proposing this system for dealing with potential Cassandras and their warnings, we want to revisit the issue of complexity, which we believe will increasingly play a role in twenty-first-century problems. Complexity was the leitmotiv of the latter half of this book. Contemporary dangers are often hidden in complexity. Because the problems, their solutions, or both, stem from rapidly advancing technology or knowledge, decision makers may not understand fully the details of what Cassandras tell them. Consequently, decision makers are increasingly reliant on technical experts, who themselves are increasingly specialized in narrow areas. Yet detailed knowledge of several different fields must often be integrated to solve complex problems.

Complexity hides vulnerabilities, creating new problems or complicating existing ones. Nowhere do we believe this tendency is clearer than in the inevitable convergence of two fields: artificial intelligence and the Internet of Things. Artificial intelligence programs running computer networks that control telecommunications, commerce, and infrastructure will increasingly "talk" with one another, resulting in unanticipated interactions and unexpected results. These networks will communicate with your car, your home electronics, even your medical devices. The initial algorithms are created by humans, but eventually lines of code will be written autonomously by software programs: code writing code. This ever more complex ecosystem will include deeply buried errors, inefficiencies, and vulnerabilities.

Coding errors are no longer a problem just for computer software. Edits to the human genetic code using CRISPR/Cas9, perhaps to remove disease-causing defects, perhaps to enhance physical attributes, may have unintended consequences years or even generations later. We do not yet fully understand how genes are expressed and interact with one another. Maybe we never will. Some scientists claim that we can simply undo problematic gene edits with the same technique that put them there in the first place. Others are not so sure it will be that easy.

We are not confronting the problem of complexity for the first time. Organizations conducting critical yet risky operations have already developed some ways to deal with such complex systems. These so-called "High Reliability Organizations," such as those involving human space flight and nuclear reactors, require extensive documentation, simulation, testing, change-management protocols, red-teaming, fault-tree analysis, and intentional redundancy. These measures can and should be adapted to deal with the rise in complexity occurring in genomics, artificial intelligence, climate science, weapon systems, health care, finance, and critical infrastructure.

This complexity issue also strengthens our belief in the need for a formal system to listen for and evaluate Cassandra's warnings. The processes used by High Reliability Organizations increase both the expense and time necessary for deployment of a new technology, but market competition demands low-cost, first-to-market solutions, not high-reliability ones. Future Cassandras will be those who see risk in complexity and advocate employing the techniques of High Reliability Organizations to ensure that we understand the implications and vulnerabilities, unavoidably slowing time to market.

THE VALUE OF LISTENING

We end this journey, at least for now, hoping that you and others will take the time to listen for contemporary Cassandras. It is a difficult and sometimes contentious process, as we have seen, but we urge you to consider establishing the ethos, processes, and teams to listen for the inconvenient warning that may be prescient.

As we said at the start of this chapter, we are optimists. We believe that the recent decades of scientific and engineering breakthroughs place us on the cusp of a brave new world, one that promises great improvement for humanity. To achieve that elevated status, however, we must be on the lookout for the potential concomitant catastrophes that would slow or reverse that progress, as continued progress and improvement are not inevitable. Thus, we must systematically identify the people who see the risks first, test what these potential Cassandras are saying, then make transparent and explicit decisions about how to deal with the risk.

This is a demanding and never-ending process, but we believe that doing it would raise the quality of decision making within corporations, governments, and the global community. Given the potential risks we will be facing in this century, the costs of not facing them in time will be unprecedented. Thus, it is important to take the time to listen for Cassandra. Can you hear her?

ACKNOWLEDGMENTS

We wish to thank the Cassandras profiled in this volume for their time and their willingness to help us understand their concerns, their lives, and them. They show not only perspicacity but also courage.

Our research team members were the sine qua non of the effort to create this volume. Chief among them, and our collaborator from day one, was Andrew S. Janiszewski, a decorated Iraq War veteran, superb intelligence analyst, and creative intellect, now at the University of Chicago's Booth School.

Likewise, David Georgi, Christen Hadfield, Alan Huffman, and Becca Worby, noted writers themselves, brought tremendous skill and intellect to this volume. Thank you.

The man who helped us understand climate change and sea-level rise was Peter Oesterling Jr., from the Korbel School at the University of Denver. Peter is now on a Fulbright Scholarship in the Peruvian Andes, studying climate change effects on South American glaciers.

Key to the accurate and timely production of this volume was Tyler Pedigo of Johns Hopkins University's Nitze School. We thank him for his diligence.

The thoughtful and expert insights of Randy Eddy, Dr. Charles Fletcher, David Cohen, Jane Geniesse, Mark Gerson, Mike Hopmeier, Mike Mears, Auren Hoffman, and numerous members of

Dialog made this book much stronger and allowed us to better understand and describe many of these complicated topics.

Udi Chattopadhyay, Kate Crumrine, Todd Egeland, Matthew Kelleher, Matthew Moneyhon, Ned Peterson, Evan Pressman, and the entire Ergo team as well as Jim Allen, Graham Allen, and Todd Asmuth, have been looking for and thinking about Cassandras since this project began. Thank you.

Finally, we extend a genuine appreciation to Dan Halperin for having faith in us, and to Zach Wagman and the entire Ecco team for making the creation of *Warnings* such a pleasure.

NOTES

CHAPTER 1: CASSANDRA: FROM MYTH TO REALITY

1. The U.S. military calls their decision-making process the "OODA loop": observe, orient, decide, act. The challenge of identifying Cassandras is one of first properly observing and then wisely deciding.
2. William Manchester, *The Last Lion: Winston Spencer Churchill: Alone, 1932–1940* (London: Pan Macmillan, 2015).

CHAPTER 2: THE SPOOK: INVASION OF KUWAIT

1. Interview with Charlie Allen, August 4, 2015.
2. Daniel L. Schacter et al., *Psychology*, 3rd ed. (New York: Worth, 2014).

CHAPTER 3: THE REBUILDER: HURRICANE KATRINA

1. Ivor van Heerden, *The Storm: What Went Wrong and Why During Hurricane Katrina* (New York: Penguin, 2007), 78–81; U.S. House of Representatives, *A Failure of Initiative: Final Report of the Select Bipartisan Committee to Investigate the Preparation for and Response to Hurricane Katrina*, Feb. 15, 2006, p. 81; and Madhu Beriwal, "Preparing for a Catastrophe: The Hurricane Pam Exercise," Statement before the Senate Homeland Security and Governmental Affairs Committee, Jan. 24, 2006.
2. Douglas Brinkley, *The Great Deluge: Hurricane Katrina, New Orleans, and the Mississippi Gulf Coast* (New York: HarperCollins, 2006), 5–7.
3. John McQuaid and Mark Schleifstein, "Special Report: Washing Away," *New Orleans Times-Picayune*, June 23–27, 2002.
4. "Flood History of Mississippi," National Oceanographic and Atmospheric Administration, Southern Regional Headquarters, http://webcache.googleusercontent.com/search?q=cache:gV29U-

Llq90J:www.srh.noaa.gov/media/jan/hydro/flood_history_ms.pdf+&c
d=1&hl=en&ct=clnk&gl=us (accessed Nov. 9, 2016).

5. "Mississippi Rivers and Tributaries Project," U.S. Army Corps of
Engineers, Mississippi Valley Division, www.mvd.usace.army.mil/
About/Mississippi-River-Commission-MRC/Mississippi-River-
Tributaries-Project-MR-T (accessed Oct. 2, 2015).

6. Richard Campanella, "Above-Sea-Level New Orleans: The
Residential Capacity of Orleans Parish's Higher Ground," Center for
Bioenvironmental Research, 2007, http://richcampanella.com/assets/
pdf/study_Campanella%20analysis%20on%20Above-Sea-Level%20
New%20Orleans.pdf (accessed Aug. 13, 2015).

7. U.S. Energy Information Administration, "Crude Oil Production"
and "Natural Gas Gross Withdrawals and Production," July 31, 2015.

8. Interview with Dr. Ivor van Heerden, Aug. 24, 2015. (We will not
repeat the footnote for other texts and quotes from interviews with van
Heerden that permeate this chapter.)

9. Brett Martel, "If the Levees Break, New Orleans Is Sunk," *Pittsburgh
Tribune Review*, Sept. 15, 2004; and van Heerden, *Storm*, 5–7.

10. R. B. Seed et al., *Investigation of the Performance of the New Orleans
Flood Protection Systems in Hurricane Katrina on August 29, 2005*, vol.
1, Independent Levee Investigation Team, July 31, 2006, 2–3 to 2–12.

11. Jon Nordheimer, "Nothing's Easy for New Orleans Flood Control,"
New York Times, Apr. 30, 2002.

12. U.S. Senate, *Hurricane Katrina: A Nation Still Unprepared*, Special
Report of the Committee on Homeland Security and Governmental
Affairs (Washington: 2006), 109; and Henry B. Hogue and Keith
Bea, "Federal Emergency Management and Homeland Security
Organization: Historical Developments and Legislative Options,"
(Washington, DC: Congressional Research Service, June 1, 2006),
18–25.

13. "Chertoff: Katrina Scenario Did Not Exist," CNN, Sept. 5, 2005,
www.cnn.com/2005/US/09/03/katrina.chertoff.

14. Ivor van Heerden, G. Paul Kemp, et al., *The Failure of the New
Orleans Levee System during Hurricane Katrina* (Baton Rouge:
Louisiana Department of Transportation and Development,
Louisiana State Project No. 704–92–0022, Dec. 18, 2006), 96–112.

15. Mark Schleifstein, "New Orleans Area Hurricane Levee Designs
Should Be Reanalyzed by 2018, Corps of Engineers Says," NOLA
.com, Apr. 20, 2015, www.nola.com/environment/index.ssf/2015/01/
new_orleans_area_hurricane_lev.html (accessed Oct. 3, 2016).

16. Mark Schleifstein, "LSU Spent Nearly $1 Million on Legal Fight
over Firing of Coastal Researcher Ivor van Heerden," NOLA.com,
Apr. 2, 2013, www.nola.com/environment/index.ssf/2013/04/lsu_
spent_nearly_1_million_on.html (accessed Oct. 3, 2016).

CHAPTER 4: THE ARABIST: THE RISE OF ISIS

1. Fawaz A. Gerges, *ISIS: A History* (Princeton, NJ: Princeton University Press, 2016), 98–128.
2. In the spirit of full disclosure, we state that both authors of this book serve on the Board of Governors of the institute, which is a nonprofit educational organization.
3. Interview with Robert Ford, September 8, 2015.

CHAPTER 5: THE SEISMOLOGIST: FUKUSHIMA NUCLEAR DISASTER

1. Sherpem Sherpa, "Japan Expert Says Atomic Warnings Ignored," DW (Deutsche Welle), www.dw.com/en/japan-expert-says-atomic-warnings-ignored/a-6488078 (accessed Oct. 3, 2016).
2. Mark Willacy, *Fukushima: Japan's Tsunami and the Inside Story of the Nuclear Meltdowns* (Sydney: Macmillan Australia, 2013).
3. Meeting minutes as quoted in *Fukushima* by Mark Willacy.
4. Ibid.
5. Interview with Dr. Yukinobu Okamura, May 27, 2016.
6. The famous physicist and creator of the first nuclear reactor, Enrico Fermi, proved to be a Cassandra about the danger of tickling the dragon's tail. After witnessing one of these haphazard tests, he told the Los Alamos scientists they would be "be dead within the year" if they continued the dangerous practice.
7. It was the same pit of plutonium that killed both men. Previously named Rufus and slated to be used for a nuclear test over Bikini Atoll, it was then renamed and melted down into other pits.
8. "Recovery and Reconstruction from the Great East Japan Earthquake," Reconstruction Agency, Tokyo, www.reconstruction.go.jp/english/topics/Progress_to_date/index.html (accessed Oct. 3, 2016).
9. Ishibashi Katsuhiko, "Why Worry? Japan's Nuclear Plants at Grave Risk from Quake Damage," *International Herald Tribune*, Aug. 11, 2007.
10. Tokyo Electric Power Company Inc., "Fukushima Nuclear Accident Summary & Nuclear Safety Reform Plan," Mar. 29, 2013, http://www.nrmc.jp/en/report/detail/__icsFiles/afieldfile/2014/04/29/report_3_4_e.pdf (accessed Jan. 30, 2017).
11. Willacy, *Fukushima*.
12. Ibid.
13. Amory Lovins, *Soft Energy Paths for the 21st Century* (Rocky Mountain Institute), 2011, http://www.rmi.org/Knowledge-Center/Library/2011-09_GaikoSoftEnergyPaths.
14. David Nakamura and Chico Harlan, "Japanese Nuclear Plant's Evaluators Cast Aside Threat of Tsunami," *Washington Post*, Mar.

23, 2011, www.washingtonpost.com/world/japanese-nuclear-plants-evaluators-cast-aside-threat-of-tsunami/2011/03/22/AB7Rf2KB_story.html (accessed Oct. 4, 2016).

CHAPTER 6: THE ACCOUNTANT: MADOFF'S PONZI SCHEME

1. Enormous amounts have been written about the Madoff case, but we benefited particularly from Harry Markopolos's own book, *No One Would Listen: A True Financial Thriller* (Hoboken, NJ: Wiley, 2010); Erin Arvedlund, *Too Good To Be True: The Rise and Fall of Bernie Madoff* (New York: Portfolio, 2009); U.S. Security and Exchange Commission Office of Inspector General, *Investigation of Failure of the SEC to Uncover Bernard Madoff's Ponzi Scheme (Public Version)* (2009); and a series of articles by Mark Seal that appeared in *Vanity Fair* magazine as "The Madoff Chronicles," in April, June, and September 2009.
2. Gregory Zuckerman and Kara Scannell, "Madoff Misled SEC in 2006, Got Off," *Wall Street Journal*, Dec. 18, 2009.
3. Interview with Harry Markopolos, Apr. 13, 2016. (We will not repeat the footnote for other texts and quotes from interviews with Markopolos that permeate this chapter.)
4. Markopolos, *No One Would Listen*, 113.
5. Ibid., 55.
6. Ibid., 62.
7. SEC Northeast Regional Office assistant regional director of enforcement Leslie Kazon, Apr. 5, 2001, e-mail to NERO regional director for enforcement Mark Schonfeld, quoted in the SEC Office of Inspector General, *Investigation of Failure of the SEC to Uncover Bernard Madoff's Ponzi Scheme*, 71.
8. Diana B. Henriques, "Madoff Scheme Kept Rippling Outward, Crossing Borders," *New York Times*, Dec. 20, 2008; and Vicky Ward, "Greenwich Mean Time," *Vanity Fair*, Apr. 2009, pp. 102–7.
9. Steve Fishman, "The Monster Mensch," *New York*, Feb. 22, 2009.
10. See Markopolos, *No One Would Listen*, 98–106.
11. Ibid., 128–29.
12. Markopolos, *No One Would Listen*, 168.
13. Gregory Zuckerman and Kara Scannell, "Madoff Misled SEC in '06, Got Off," *Wall Street Journal*, Dec. 18, 2008.
14. Ibid., 192.
15. Harry Markopolos e-mail to Meaghan Cheung, June 2007, quoted in ibid.
16. *U.S. v. Bernard L. Madoff*, FBI Special Agent Ted Cacioppi, in his Complaint in U.S. Federal Court, Southern District of New York, Dec. 11, 2008.

17. Ibid.
18. Arvedlund, *Too Good to Be True*, 231.
19. Portfolio Staff, "Wiesel Lost 'Everything' to Madoff," *Upstart Business Journal*, Feb. 26, 2009, http://upstart.bizjournals.com/executives/2009/02/26/Elie-Wiesel-and-Bernard-Madoff.html?page=all.
20. M. J. Lee, "Madoff: Politics, Remorse, Wall Street," Politico.com, Mar. 20, 2014, www.politico.com/story/2014/03/bernie-madoff-interview-104838 (accessed Nov. 10, 2016).
21. Markopolos, *No One Would Listen*, 127.
22. Ibid.
23. SEC's *Investigation of Failure* report (2009), 37.
24. Ibid., 24.
25. Ibid., 261.
26. SEC's *Investigation of Failure* report (2009), 37.
27. Deborah Solomon, "Math Is Hard," *New York Times*, Feb. 25, 2010, http://www.nytimes.com/2010/02/28/magazine/28fob-q4-t.html (accessed Oct. 25, 2016).
28. Markopolos, *No One Would Listen*, 226.

CHAPTER 7: THE INSPECTOR: MINE DISASTER

1. Jane R. Eggleston, "History of West Virginia Mineral Industries—Coal," West Virginia Geological and Economic Survey, Department of Commerce, Sept. 1996, www.wvgs.wvnet.edu/www/geology/geoldvco.htm (accessed Oct. 4, 2016).
2. Sean Patrick Adams, "The U.S. Coal Industry in the Nineteenth Century," Economic History Association, https://eh.net/encyclopedia/the-us-coal-industry-in-the-nineteenth-century-2 (accessed Oct. 4, 2016); and John A. Breslin, *One Hundred Years of Federal Mining Safety and Health Research*, U.S. Department of Health and Human Services, Centers for Disease Control and Prevention, National Institute for Occupational Safety and Health, No. 2010–128 (Pittsburgh: NIOSH, Feb. 2010), 1–2.
3. Data derived from "Coal Mining Disasters: 1839 to Present," U.S. Department of Health and Human Services, Centers for Disease Control and Prevention, National Institute for Occupational Safety and Health, Jan. 2016, www.cdc.gov/niosh/mining/statistics/content/coaldisasters.html (accessed Oct. 4, 2016).
4. J. Davitt McAteer, *Monongah: The Tragic Story of the 1907 Monongah Mine Disaster* (Morgantown: University of West Virginia Press, 2007), 253.
5. Ibid., 242–58l; and interview with J. Davitt McAteer, Apr. 15, 2016.

(We will not repeat the footnote for other texts and quotes from interviews with McAteer that permeate this chapter.)

6. Derived from NIOSH data.

7. Here we use the same definition of "mining disaster" as the Mine Safety and Health Administration, which is an incident that results in the death of five or more.

8. Public Law 77–49, U.S. Government Printing Office, May 7, 1941.

9. Robert E. Hartley, *Death Underground: The Centralia and West Frankfort Mine Disasters*, 1st ed. (Carbondale: Southern Illinois University Press, July 24, 2006), 46.

10. Public Law 552, July 16, 1952, 66 Stat. 877, U.S. Government Publishing Office, https://www.gpo.gov/fdsys/pkg/STATUTE-66/pdf/STATUTE-66-Pg692.pdf (accessed Oct. 4, 2016).

11. Breslin, *One Hundred Years of Federal Mining Safety and Health Research*, 39–41.

12. Derived from NIOSH data.

13. Rachel Scott, "'Yes, Sir, This Has Certainly Been Considered a Safe Mine,'" *Atlantic*, Dec. 1972, www.theatlantic.com/magazine/archive/1972/12/-yes-sir-this-has-certainly-been-considered-a-safe-mine/304565 (accessed Oct. 5, 2016).

14. Derived from NIOSH data.

15. Laurent Belsie, "West Virginia Mine Explosion: Mining Jobs Were Getting Safer, Until Now," *Christian Science Monitor*, Apr. 6, 2010, www.csmonitor.com/Business/new-economy/2010/0406/West-Virginia-mine-explosion-Mining-jobs-were-getting-safer-until-now (accessed Oct. 5, 2016).

16. J. Davitt McAteer et al., *Upper Big Branch: The April 5, 2010, Explosion—a Failure of Basic Coal Mine Safety Practices*, (Charleston, WV: Governor's Independent Investigation Panel, May 2011), 23–25.

17. "How Fossil Fuels Were Formed," US Department of Energy, Fossil Energy Office of Communications, Feb. 12, 2013, www.fe.doe.gov/education/energylessons/coal/gen_howformed.html (accessed Oct. 5, 2016).

18. A brief but helpful video for those unfamiliar with longwall mining operations is posted to YouTube by Caterpillar Global Mining, available at https://youtube/bXORrVmxwbM (accessed Oct. 5, 2016).

19. Fred Kissell, Jerry C. J. Tien, and Edward D. Thimons, "Methods for Controlling Explosion Risk at Coal Mine Working Faces," Proceedings of the 32nd International Conference of Safety in Mines Research Institutes, Sept. 28–29, 2007, Beijing, China, available at https://www.cdc.gov/niosh/mining/UserFiles/works/pdfs/mfcer.pdf (accessed Oct. 5, 2016).

20. Alvin L. Brown et al., *Report of Investigation: Fatal Underground*

Mine Explosion, April 5, 2010 (Arlington, VA: U.S. Department of Labor, Mine Safety and Health Administration, 2011), 3.

21. "Transcript of the Testimony of Joseph Mackowiak," Sargent's Court Reporting Services, May 17, 2010, pp. 12–15.

22. McAteer et al., *Upper Big Branch*, 54.

23. Dennis B. Roddy and Vivian Nereim, "A History of Violations at Upper Big Branch Mine," *Pittsburgh Post-Gazette*, Apr. 6, 2010, www.post-gazette.com/region/2010/04/06/A-history-of-violations-at-Upper-Big-Branch-mine/stories/201004060199 (accessed Oct. 5, 2016).

24. Mackowiak Testimony, May 18, 2010, pp. 25–26; and Ken Ward Jr., "Emails: MSHA Asked Massey Exec to Send Ventilation Expert to Fix Airflow Prior to Upper Big Branch Blast," *Charleston Gazette-Mail*, Coal Tattoo Blog, May 11, 2011, http://blogs.wvgazettemail.com/coaltattoo/2011/05/11/emails-msha-asked-massey-exec-to-send-ventilation-expert-to-fix-airflow-prior-to-upper-big-branch-blast/comment-page-1 (accessed Oct. 5, 2016).

25. Brown et al., *Report of Investigation: Fatal Underground Mine Explosion, April 5, 2010*; and "Upper Big Branch–South Explosion Scenario," U.S. Department of Labor, Mine Safety and Health Administration, Jan. 10. 2014, YouTube, https://youtube/hBvaJVaAEDU (accessed Oct. 5, 2016).

26. Sarah Tincher, "Bill Ross Recalls Efforts to Make Massey's Mines Safer in Day 23 of Blankenship Trial," *Charleston State Journal*, Nov. 4, 2015, www.statejournal.com/story/30435688/bill-ross-recalls-efforts-to-make-masseys-mines-safer-in-day-23-of-blankenship-trial (accessed Oct. 5, 2016).

27. Bill Chappell, "Former Coal CEO Blankenship Found Guilty of Conspiracy in Mine Disaster Case," National Public Radio, Dec. 3, 2015, www.npr.org/sections/thetwo-way/2015/12/03/456533421/former-coal-ceo-blankenship-found-guilty-of-conspiracy-in-mine-disaster-case (accessed Oct. 5, 2016).

28. *Internal Review of MSHA's Actions at the Upper Big Branch Mine-South*, U.S. Department of Labor, Mine Safety and Health Administration, Mar. 6, 2012.

CHAPTER 8: THE MARKET ANALYST: THE 2008 RECESSION

1. Interview with Meredith Whitney, June 3, 2016.

2. S&P, Moody's, and Fitch were the largest such rating agencies for these collections of mortgages.

3. "Warren Buffet on Derivatives," edited excerpts from Berkshire Hathaway 2002 Annual Report, www.fintools.com/docs/Warren%20Buffet%20on%20Derivatives.pdf (accessed Oct. 5, 2016).

4. Sheelah Kolhatkar's piece, "In Meredith Whitney We Trust?" was an important source for this chapter and used extensively. *New York*, Mar. 22, 2009, http://nymag.com/news/businessfinance/55497 (accessed Oct. 5, 2016).

5. Steve Rosenbush, "The Analyst Who Rocked Citi," Bloomberg, Nov. 27, 2007. www.bloomberg.com/news/articles/2007–11–26/the-analyst-who-rocked-citibusinessweek-business-news-stock-market-and-financial-advice (accessed Oct. 6, 2016).

6. Ibid.

7. Meredith Whitney, *Fate of the States: The New Geography of American Prosperity* (New York: Portfolio, 2013).

8. Kolhatkar, "In Meredith Whitney We Trust?"

9. Ibid.

10. William Cohan, *House of Cards: A Tale of Hubris and Wretched Excess on Wall Street* (New York: Anchor, 2010), 47.

11. Kolhatkar, "In Meredith Whitney We Trust?"

12. Michael Lewis, "California and Bust," *Vanity Fair*, Nov. 2011, www.vanityfair.com/news/2011/11/michael-lewis-201111 (accessed Nov. 9, 2016).

CHAPTER 9: THE CASSANDRA COEFFICIENT

1. See Andrew C. Revkin, "'Disaster Memory' and the Flooding of Fukushima," Dot Earth blog, *New York Times*, Apr. 4, 2011, http://dotearth.blogs.nytimes.com/2011/04/04/disaster-memory-and-the-flooding-of-fukushima (accessed Oct. 6, 2016).

2. See Wim De Neys and Jean-François Bonnefon, "The 'Whys' and 'Whens' of Individual Differences in Thinking Biases," *Trends in Cognitive Sciences* 17, no. 4 (2013): 172–78.

3. "Michelin Lecture," California Institute of Technology, January 17, 2003.

4. Trafton Drew, Melissa L. H. Vo, and Jeremy M. Wolfe, "The Invisible Gorilla Strikes Again: Sustained Inattentional Blindness in Expert Observers," *Psychological Science* 24, no. 9 (Sept. 2013): 1848–53, www.ncbi.nlm.nih.gov/pmc/articles/PMC3964612 (accessed Oct. 7, 2016).

5. See Richard Farson, *Management of the Absurd* (New York: Simon & Schuster, 1996).

6. As discussed in Eliot A. Cohen and John Gooch's text *Military Misfortunes: The Anatomy of Failure in War* (New York: Free Press, 2006) and their assessment of three components of failure: failure to learn, failure to anticipate, and failure to adapt.

7. See Tobias Greitemeyer and Dirk Oliver Mügge, "When Bystanders

Increase Rather Than Decrease Intentions to Help," *Social Psychology* 46 (2015): 116–19, http://econtent.hogrefe.com/doi/abs/10.1027/1864-9335/a000215?journalCode=zsp& (accessed Oct. 7, 2016).

8. This is an effect of what Micha Zenko in his book *Red Team* terms "grading your own homework" (p. xvi). Organizations will rationalize reasons not to change based on the aforementioned groupthink schema deeply embedded in human psychology. Habit is the enemy in this instance. While it is considered a survival mechanism to defend against dimensionality reduction, or perceived loss, humans, in all of their clever rationalizing repeatedly fail to accurately assess their own level of effectiveness. That's why "red teams," or "devil's advocate" associates are useful in the process of assessing weak links in the operational chain. See Micha Zenko, *Red Team: How to Succeed by Thinking Like the Enemy* (New York: Basic Books, 2015).

9. Chris Walsh, "The Soul of Cowardice," Opinionator blog, *New York Times*, Nov. 9, 2014, http://opinionator.blogs.nytimes.com/2014/11/09/the-soul-of-cowardice/?_r=0 (accessed Oct. 7, 2016).

10. On satisficing, see Marcel Zeelenberg, "Robust Satisficing via Regret Minimization," *Journal of Marketing Behavior* 1, no. 2 (Oct. 2015): 157–66 https://www.researchgate.net/profile/Marcel_Zeelenberg/publication/283278882_Robust_Satisficing_via_Regret_Minimization/links/5631ecb708ae506cea679fa3.pdf (accessed Oct. 7, 2016).

11. Tomas Chamorro-Premuzic, "The Five Characteristics of Successful Innovators," *Harvard Business Review*, Oct. 25, 2013, https://hbr.org/2013/10/the-five-characteristics-of-successful-innovators (accessed Oct. 7, 2016).

12. This could be a direct result of social projection that refuses to consider and adopt an alternative hypothesis for "normative" behavior due to the adoption "of an overly restrictive hypothesis about what the 'rule' is" (Keith A. Stanovich, *What Intelligence Tests Miss: The Psychology of Rational Thought* [New Haven, CT: Yale University Press], 143). The "failure to adapt" (Cohen and Gooch, *Military Misfortunes*, 133–164) is not on the part of Cassandra; it's a result of the "inadequate or imperfect anticipation of an enemy's actions" wherein Cassandra may be mistaken for the enemy.

13. Jeremy D. Coplan et al., "The Relationship between Intelligence and Anxiety: An Association with Subcortical White Matter Metabolism," *Frontiers in Evolutionary Neuroscience* 3, no. 8 (2011), www.ncbi.nlm.nih.gov/pmc/articles/PMC3269637 (accessed Oct. 7, 2016).

14. In fact, "while excessive worry is generally seen as a negative trait and high intelligence as a positive one, worry may cause our species

to avoid dangerous situations, regardless of how remote a possibility they may be . . . worry may make people 'take no chances,' and such people may have higher survival rates. Thus, like intelligence, worry may confer a benefit upon the species. In Israel Dr. Tsachi Ein-Dor drew a similar conclusion, that people with higher anxiety levels tend to detect threats sooner and warn others.

15. In spite of the persistent view that science is the ultimate model of objectivity, scientists are, after all, human beings and still subject to what philosopher Nick Bostrom terms the selection effect. The scientists and their instruments reflect the fact that "all observations require the existence of an appropriately positioned observer." See Nick Bostrom, *Anthropic Bias: Observation Selection Effects in Science and Philosophy* (New York: Routledge, 2002).

16. Charles S. Taber and Milton Lodge, "Motivated Skepticism in the Evaluation of Political Beliefs," *American Journal of Political Science* 50, no. 3 (2006): 755–69, doi:10.1111/j.1540–5907.2006.00214.

17. Authority bias is perhaps best shown in the classic 1961 Milgram experiment in which volunteers were convinced to "shock" other volunteers because an authority figure encouraged them to do so.

18. See Yuval Noah Harari, *Sapiens: A Brief History of Humankind* (New York: HarperCollins, 2014) for a fascinating review of human cognitive evolution and how it leaves us riddled with decision-making biases.

19. Having data also does not make one infallible. Bostrom: "What these various topics have in common is that they involve the assignment of conditional credences of the form 'I make such and such observations given that the world is such and such,'" thus establishing that the observer is interpreting data based on a preconception of the world as it "is" or "should be" (p. 4). Nick Bostrom, "Self-Locating Belief in Big Worlds: Cosmology's Missing Link to Observation," *Journal of Philosophy* 99, no. 12 (2002): 607–23.

20. Richard E. Neustadt and Ernest R. May, *Thinking in Time: The Uses of History for Decision Makers* (New York: Free Press, 1986), 48–56.

21. The subject of imperfect recall and how it affects both memory and reaction constraints is relevant here. Michele Piccone and Ariel Rubenstein elaborate on the interpretation of actionable models based upon limitation of the decision maker's data set, whereby "the decision maker cannot plan to assign different actions to two histories which lie in the same information set" (p. 4). This is an attribute of dimensionality reduction via overfitting the data to match the imperfect recall of a prior seemingly similar experience. See Michele Piccone and Ariel Rubinstein, "On the Interpretation of Decision Problems with Imperfect Recall," *Games and Economic Behavior* 20 (1997): 3–24.

22. Also known as the "no man in the dock" theory, wherein failure is seen as not predicated on a single individual but distributed among those making "command decisions," along with "organizational deficiencies and changes in the environment" (Cohen and Gooch, *Military Misfortunes*, 6).

CHAPTER 10: THE COMPUTER SCIENTIST: ARTIFICIAL INTELLIGENCE

1. Brandon Vinson "X-47B Makes First Arrested Landing at Sea," U.S. Navy at http://www.navy.mil/submit/display.asp?story_id=75298, accessed July 7 2013.
2. Just as artificial intelligence is an overbroad term, so are the terms used to refer to its two constituent halves, weak and strong. Weak AI is often also called narrow AI.
3. Arthur Samuel offered this definition of machine learning in 1959.
4. Superintelligence is a term made popular by philosopher Nick Bostrom and is often also called artificial general intelligence.
5. Referenced from Luke Muehlhauser's fantastic work which was a great guide for the authors.
6. James Barrat, *Our Final Invention: Artificial Intelligence and the End of the Human Era* (New York: Thomas Dunne Books, 2013). Barrat's book was an important source for the authors.
7. Nick Bostrom, "Ethical Issues in Advanced Artificial Intelligence," http://www.nickbostrom.com/ethics/ai.html (accessed Nov. 9, 2016). Bostrom is a leading thinker on AI as well as disasters. His body of research was highly valuable to the authors.
8. MIT Aeronautics and Astronautics Department Centennial Symposium, Oct. 2014, http://aeroastro.mit.edu/aeroastro100/centennial-symposium (accessed Oct. 8, 2016).
9. Question 5 in "Less Wrong Q&A with Eliezer Yudkowsky," Jan. 2010, http://lesswrong.com/lw/1lq/less_wrong_qa_with_eliezer_yudkowsky_video_answers (accessed Oct. 8, 2016).
10. Interview with Eliezer Yudkowsky, March 15, 2016.
11. Thanks to Luke Muehlhauser, "Facing the Intelligence Explosion," http://intelligenceexplosion.com/en/2011/not-built-to-think-about-ai; and Martin Kaste, "The Singularity: Humanity's Last Invention?" *All Things Considered*, National Public Radio, Jan. 11, 2011, www.npr.org/2011/01/11/132840775/The-Singularity-Humanitys-Last-Invention (accessed Oct. 8, 2016).
12. MIRI Mission Statement.
13. Yes, we asked Yudkowsky how he convinced the gatekeepers to let him out, but he's keeping that to himself.
14. Eliezer Yudkowsky, "Lonely Dissent," Less Wrong, Dec. 28, 2007,

http://lesswrong.com/lw/mb/lonely_dissent (accessed Oct. 8, 2016).

15. David Gilbert, "From Deep Mind to Watson: Why You Should Stop Worrying and Love AI," *International Business Times*, Mar. 18, 2016, www.ibtimes.com/deepmind-watson-why-you-should-learn-stop-worrying-love-ai-2339231 (accessed Oct. 8, 2016), quoting Harriet Green, general manager of Watson's Internet of Things Unit.

16. Ray Kurzweil, "Don't Fear Artificial Intelligence," www.kurzweilai .net/dont-fear-artificial-intelligence-by-ray-kurzweil (accessed Oct. 8, 2016).

17. Vincent C. Müller and Nick Bostrom, "Future Progress in Artificial Intelligence: A Survey of Expert Opinion," in Vincent C. Müller, ed., *Fundamental Issues of Artificial Intelligence* (Berlin: Springer, Synthese Library, 2014), available online at www.nickbostrom.com/ papers/survey.pdf (accessed Oct. 8, 2016).

18. Ibid.

19. DARPA is also the creative force behind the Internet, automated voice recognition, stealth technology, and more.

20. DARPA deputy director Steven Walker quoted in Sydney J. Freedberg Jr., "Robots, Techies, and Troops: Carter and Roper on 3rd Offset," Breaking Defense, June 13, 2016, http://breakingdefense .com/2016/06/trust-robots-tech-industry-troops-carter-roper (accessed Oct. 8, 2016).

21. Michael Sainato, "Steven Hawking, Elon Musk, and Bill Gates Warn About Artificial Intelligence," *The Observer* (UK), Aug. 19, 2015, http://observer.com/2015/08/stephen-hawking-elon-musk-and-bill-gates-warn-about-artificial-intelligence (accessed Oct. 8, 2016); and Elon Musk interview with MIT students at the MIT Aeronautics and Astronautics Department Centennial Symposium, Oct. 2014, http:// aeroastro.mit.edu/aeroastro100/centennial-symposium (accessed Oct. 8, 2016).

22. Bloomberg via Shobhit Seth, "The World of High Frequency Algorithmic Trading," Investopedia, Sept. 16, 2015, www .investopedia.com/articles/investing/091615/world-high-frequency-algorithmic-trading.asp (accessed Oct. 8, 2016).

23. Andrew Ng, "Is A.I. an Existential Threat to Humanity?" Quora, https://www.quora.com/Is-AI-an-existential-threat-to-humanity/answer/ Andrew-Ng (accessed Oct. 8, 2016).

24. The study looks at jobs at risk from weak AI and robotics. Carl Benedikt Frey and Michael A. Osbourne, "The Future of Employment: How Susceptible Are Jobs to Computerisation?" Sept. 17, 2013, Oxford Martin School, www.oxfordmartin.ox.ac.uk/ downloads/academic/The_Future_of_Employment.pdf (accessed Oct. 8, 2016).

25. Nicholas Carr, *The Glass Cage: Automation and Us* (New York: Norton, 2014), reviewed in Sean Braswell, "All Rise for Chief Justice Robot!" Ozy.com, www.ozy.com/immodest-proposal/all-rise-for-chief-justice-robot/41131 (accessed Oct. 8, 2016).

26. McKinsey Global Institute, referenced in Lakshmi Sandhana, "47% of U.S. Jobs under Threat from Computerization According to Oxford Study," New Atlas, http://newatlas.com/half-of-us-jobs-computerized/29142 (accessed Oct. 8, 2016).

27. John Maynard Keynes, "Economic Possibilities for Our Grandchildren" (1933) in *Essays in Persuasion* (New York: Norton, 1963) referenced in Frey and Osbourne, "Future of Employment," 3.

28. Studebaker was a leading buggy manufacturer that shifted to cars and became one of the four largest auto manufacturers in the United States.

29. Martin Ford, *Rise of the Robots: Technology and the Threat of a Jobless Future* (New York: Basic Books, 2015).

30. "The Great Depression and World War II, 1929–1945," Gilder Lehrman Institute of American History, www.gilderlehrman.org/history-by-era/essays/great-depression-and-world-war-ii-1929–1945 (by subscription, free to students and teachers).

31. Ibid.

32. Ash Carter, Secretary of Defense Testimony to House Appropriations Committee, Feb. 25, 2016, www.defense.gov/News/Speeches/Speech-View/Article/672855/submitted-statement-house-appropriations-committee-defense-fy-2017-budget-reque (accessed Oct. 8, 2016).

33. Freedberg, "Robots, Techies, and Troops."

34. William Roper, Defense One Tech Summit, June 10, 2016, in Freedberg, "Robots, Techies, and Troops."

CHAPTER 11: THE JOURNALIST: PANDEMIC DISEASE

1. Henry W. Griest, "The Big Sickness," Arctic Circle, http://arcticcircle.uconn.edu/HistoryCulture/Sickness/sickness.html (accessed Oct. 8, 2016).

2. Jeffery Taubenberger, Ann Reid, Amy Krafft, et al., "Initial Genetic Characterization of the 1918 'Spanish' Influenza Virus," *Science* 275 (1997): 1793–96.

3. David Malakoff, "Breaking News: NSABB Reverses Position on Flu Papers," American Association for the Advancement of Science, Mar. 30, 2012, www.sciencemag.org/news/2012/03/breaking-news-nsabb-reverses-position-flu-papers (accessed Nov. 9, 2016). However, not long after the wise decision to fight publication, the board reversed itself in a split vote 12–6 and decided to no longer protest.

The board's new logic was that "understanding specific mutations may improve international surveillance and public health and safety."

4. Jim Avila and Meredith Ramsey, "Renowned Bird Flu Expert Warns: Be Prepared," ABC News, Mar. 14, 2006, http://abcnews.go.com/WNT/AvianFlu/story?id=1724801&page=1 (accessed Nov. 9, 2016).

5. Robert G. Webster et al. "H5N1 Outbreaks and Enzootic Influenza," *Emerging Infectious Diseases* 12:1 (2006): 3–8.

6. Avila and Ramsey, "Renowned Bird Flu Expert Warns."

7. Interview with Dr. Robert Webster, May 23, 2016.

8. Interview with Dr. Laurie Garrett, May 23, 2016.

9. Ryan White was a hemophiliac teenager from Kokomo, Indiana, who contracted HIV from contaminated blood. In 1985, when he was fourteen, he was kicked out of middle school. He was ridiculed and received death threats. A bullet was fired through his living-room window. He died of AIDS at eighteen in 1990.

10. In 1992, when running for Senate in Arkansas, Mike Huckabee said steps should be taken to "isolate the carriers of this plague." In the 1980s, Senator Jesse Helms called for "quarantine of those infected." Dana Bash and Evan Glass, "Huckabee Refuses to Retract '92 Remarks on AIDS Patients," CNN, Dec. 10, 2007, www.cnn.com/2007/POLITICS/12/10/huckabee.aids (accessed Oct. 8, 2016); and Jim Morrison, "HIV Quarantines? They Already Exist, *Advocate*, May 15, 2013, www.advocate.com/commentary/2013/05/15/op-ed-hiv-quarantines-they-already-exist (accessed Oct. 8, 2016).

11. "Discovery and Development of Penicillin," American Chemical Society, https://www.acs.org/content/acs/en/education/whatischemistry/landmarks/flemingpenicillin.html.

12. While human misuse of antibiotics is certainly one cause of antibiotic resistance, there is a complex interplay of reasons. To take just one example, 80 percent of antibiotics are used on farm animals to increase growth.

13. World Health Organization, *Antimicrobial Resistance: Global Report on Surveillance* (Geneva: WHO, 2014), http://apps.who.int/iris/bitstream/10665/112642/1/9789241564748_eng.pdf?ua=1 (accessed Oct. 8, 2016).

14. Amy E. Smithson and Leslie-Anne Levy, *Ataxia: The Chemical and Biological Terrorism Threat and the U.S. Response*, Stimson Report 35 (Washington, DC: Stimson Center, Oct. 9, 2000), 78–79.

15. The Council on Foreign Relations is one of the oldest and most prestigious think tanks in America. It also happens to be where Garrett works.

CHAPTER 12: THE CLIMATE SCIENTIST: SEA-LEVEL RISE

1. "People," Climate Science, Awareness, and Solutions, Earth Institute, Columbia University, http://csas.ei.columbia.edu/about-us/people (accessed Oct. 9, 2016).

2. "About GISS," Goddard Institute of Space Studies, National Aeronautics and Space Administration, www.giss.nasa.gov/about (accessed Oct. 9, 2016).

3. Mark Bowen, Censoring Science: Inside the Political Attack on Dr. James Hansen and the Truth of Global Warming (New York: Dutton, 2008), 209, 211.

4. James Hansen, "Why I Must Speak Out about Climate Change, www.ted.com/talks/james_hansen_why_i_must_speak_out_about_climate_change?language=en (accessed Oct. 9, 2016).

5. James Hansen et al., "Climate Impact of Increasing Atmospheric Carbon Dioxide," Science 213, no. 4511 (1981): 957–66, at 957.

6. James E. Hansen, Storms of My Grandchildren: The Truth about the Coming Climate Catastrophe and Our Last Chance to Save Humanity (New York: Bloomsbury, 2009), 61.

7. Philip Shabecoff, "Global Warming Has Begun, Expert Tells Senate," New York Times, June 24, 1988, sec. U.S., www.nytimes .com/1988/06/24/us/global-warming-has-begun-expert-tells-senate .html (accessed Oct. 9, 2016); and Bowen, Censoring Science, 224.

8. Interview with James Hansen, Mar. 15, 2016. (We will not repeat the footnote for other texts and quotes from interviews with Hansen that permeate this chapter.)

9. Richard A. Kerr, "Hansen vs. the World on the Greenhouse Threat," Science 244, no. 4908 (1989): 1041–43, at 1041.

10. "Global Temperature," Vital Signs, NASA Global Climate Change, http://climate.nasa.gov/vital-signs/global-temperature (accessed Oct. 9, 2016).

11. Christopher B. Field, Vicente R. Barros, and Intergovernmental Panel on Climate Change, eds., Climate Change 2014: Impacts, Adaptation, and Vulnerability—Working Group II Contribution to the Fifth Assessment Report of the Intergovernmental Panel on Climate Change (New York: Cambridge University Press, 2014), 63.

12. "Adoption of the Paris Agreement," Conference of the Parties, Twenty-First Session, Paris, Nov. 30–Dec. 11, 2015, (United Nations: Framework Convention on Climate Change), 4, 6.

13. Oliver Milman, "James Hansen, Father of Climate Change Awareness, Calls Paris Talks 'a Fraud,'" Guardian, Dec. 12, 2015, sec. Environment, www.theguardian.com/environment/2015/dec/12/james-hansen-climate-change-paris-talks-fraud.

14. James Hansen et al., "Ice Melt, Sea Level Rise and Superstorms: Evidence from Paleoclimate Data, Climate Modeling, and Modern Observations That 2°C Global Warming Is Highly Dangerous," *Atmospheric Chemistry and Physics Discussions* 15, no. 14 (July 23, 2015): 20059–179, doi:10.5194/acpd-15-20059-2015, www.atmos-chem-phys.net/16/3761/2016/acp-16-3761-2016.html, p. 20061.

15. "Infographic: Sea Level Rise," NASA Global Climate Change, http://climate.nasa.gov/climate_resources/125 (accessed Oct. 9, 2016).

16. "Quick Facts on Ice Sheets," National Snow and Ice Data Center, https://nsidc.org/cryosphere/quickfacts/icesheets.html (accessed Oct. 9, 2016).

17. James Hansen, "Ice Melt, Sea Level Rise and Superstorms Video Abstract," YouTube, https://www.youtube.com/watch?v=JP-cRqCQRc8 (accessed Oct. 9, 2016).

18. Andrew C. Revkin, "Whiplash Warning When Climate Science Is Publicized Before Peer Review and Publication," Dot Earth Blog, 1437675605, http://dotearth.blogs.nytimes.com/2015/07/23/whiplash-warning-when-climate-science-is-publicized-before-peer-review-and-publication (accessed Oct. 9, 2016).

19. Hansen et al., "Ice Melt, Sea Level Rise and Superstorms," Peer Review, Interactive Discussion, *Atmospheric Chemistry and Physics*, www.atmos-chem-phys.net/16/3761/2016/acp-16-3761-2016-discussion.html#discussion (accessed Oct. 9, 2016).

20. Eli Kintisch, Hansen's Retirement from NASA Spurs Look at His Legacy," *Science* 340, no. 6132 (2013): 540–41, at 540.

21. "James Hansen: For Sounding the Alarm on Climate Change, Early and Often," *Foreign Policy* 197 (2012): 45.

22. Katherine Bagley, "For James Hansen, the Science Demands Activism on Climate," *Yale Environment 360*, http://e360.yale.edu/feature/james_hansen_science_demands_action/2981 (accessed Oct. 9, 2016).

23. Stephane Hallegatte et al., "Future Flood Losses in Major Coastal Cities," *Nature Climate Change* 3, no. 9 (Aug. 18, 2013): 802–6, doi:10.1038/nclimate1979.

24. Joanna Jolly, "How Boston Is Rethinking Its Relationship with the Sea," BBC News, www.bbc.com/news/magazine-29761274 (accessed Oct. 9, 2016).

25. "James Hansen: Taking Heat for Decades," *Bulletin of the Atomic Scientists* 69, no. 4 (July–Aug., 2013): 1–8, doi:10.1177/0096340213493256, 5.

CHAPTER 13: THE WEATHERMAN: NUCLEAR ICE AGE

1. Christine C. Fair, *Fighting to the End: The Pakistan Army's Way of War* (New York, NY: Oxford University Press, 2014), 16.

2. Bruce Riedel, *Avoiding Armageddon: America, India, and Pakistan to the Brink and Back* (Washington, DC: Brookings Institution Press, 2013).
3. Jeffrey Goldberg and Marc Ambinder, "The Ally from Hell," *The Atlantic*, Dec. 2011, www.theatlantic.com/magazine/archive/2011/12/the-ally-from-hell/308730 (accessed Nov. 11, 2016).
4. Interview with Alan Robock, May 10, 2016.

CHAPTER 14: THE ENGINEER: THE INTERNET OF EVERYTHING

1. Elinor Mills, "Joe Weiss, Crusader for Critical Infrastructure Security (Q&A)," *CNET*, May 10, 2010, www.cnet.com/news/joe-weiss-crusader-for-critical-infrastructure-security-q-a/ (accessed Nov. 9, 2016).
2. Interview with Joe Weiss, Feb. 12, 2016.

CHAPTER 15: THE PLANETARY DEFENDER: METEOR STRIKE

1. Donald K. Yeomans, *Near-Earth Objects: Finding Them Before They Find Us* (Princeton, NJ: Princeton University Press, 2013), 54–55.
2. Interview with David Morrison, Apr. 14, 2016.
3. Christopher Mele, "NASA and FEMA Rehearse for the Unthinkable: An Asteroid Strike on Los Angeles," *New York Times*, Nov. 14, 2016, http://www.nytimes.com/2016/11/15/science/nasa-and-fema-rehearse-for-the-unthinkable-an-asteroid-strike-on-los-angeles.html?_r=0 (accessed Dec. 15, 2016).

CHAPTER 16: THE BIOLOGIST: GENE EDITING

1. Interview with Jennifer Doudna, June 8, 2016. (We will not repeat the footnote for other texts and quotes from interviews with Doudna that permeate this chapter.)
2. Jennifer Kahn, "The CRISPR Quandary," *New York Times Magazine*, Nov. 9, 2015, www.nytimes.com/2015/11/15/magazine/the-crispr-quandary.html (accessed Oct. 11, 2016).
3. Andrew Pollack, "Jennifer Doudna, a Pioneer Who Helped Simplify Genome Editing," *New York Times*, May 11, 2015, www.nytimes.com/2015/05/12/science/jennifer-doudna-crispr-cas9-genetic-engineering.html?_r=0 (accessed Oct. 11, 2016).
4. Martin Jinek et al., "A Programmable Dual-RNA–Guided DNA Endonuclease in Adaptive Bacterial Immunity," *Science* 337, no. 6096 (Aug. 17, 2012): 816–21.
5. Kahn, "The CRISPR Quandary."
6. Sean B. Carroll, "Tracking the Ancestry of Corn Back 9,000 Years,"

New York Times, May 24, 2010, www.nytimes.com/2010/05/25/
science/25creature.html (accessed Oct. 11, 2016); and Jerry Adler
and Andrew Lawler, "How the Chicken Conquered the World,"
Smithsonian, June 2012, www.smithsonianmag.com/history/how-the-
chicken-conquered-the-world-87583657 (accessed Oct. 11, 2016).

7. Christina Larson, "China's Bold Push into Genetically
    Customized Animals," *Scientific American*, Nov. 17, 2015, www
    .scientificamerican.com/article/china-s-bold-push-into-genetically-
    customized-animals (accessed Oct. 11, 2016).

8. Hao Yin, Wen Xue, et al., "Genome Editing with Cas9 in Adult Mice
    Corrects a Disease Mutation and Phenotype," *Nature Biotechnology*
    32, no. 6 (Mar. 30, 2014): 551–53, DOI:10.1038/nbt.2884.

9. Chengzhu Long, Leonela Amoasii, et al., "Postnatal Genome Editing
    Partially Restores Dystrophin Expression in a Mouse Model of
    Muscular Dystrophy," *Science* 351, no. 6271 (Jan. 22, 2016): 400–
    403, DOI: 10.1126/science.aad5725.

10. Jonathan Rockoff, "Why Gene-Editing Technology Has Scientists
    Excited," *Wall Street Journal*, June 28, 2015, www.wsj.com/articles/
    why-gene-editing-technology-has-scientists-excited-1434985998
    (accessed Oct. 11, 2016).

11. Amy Maxmen, "Easy DNA Editing Will Remake the World. Buckle
    Up," *Wired*, Aug. 2015, www.wired.com/2015/07/crispr-dna-editing-2
    (accessed Oct. 11, 2016).

12. Yangfan Fu, Jennifer A. Foden, et al., "High-Frequency Off-Target
    Mutagenesis Induced by CrIsPr-Cas Nucleases in Human Cells,"
    Nature Biotechnology 31, no. 9 (Sept. 2013).

13. Helen Shen, "First Monkeys with Customized Mutations Born,"
    Nature News, Jan. 30, 2014, www.nature.com/news/first-monkeys-
    with-customized-mutations-born-1.14611 (accessed Oct. 11, 2016).

14. Paul Berg, "Meetings That Changed the world: Asilomar 1975:
    DNA Modification Secured," *Nature* 455, no. 7211 (Sept. 17, 2008):
    290–91, doi:10.1038/455290a, www.nature.com/nature/journal/v455/
    n7211/full/455290a.html (accessed Oct. 11, 2016).

15. Anthony J. F. Griffiths et al., *Making Recombinant DNA: An
    Introduction to Genetic Analysis*, 7th ed. (New York: Freeman, 2000).
    The conference was hosted by the National Center for Biotechnology
    Information.

16. The Paul Berg Papers: Recombinant DNA Technologies and
    Researchers' Responsibilities, 1973–1980, Profiles in Science,
    National Library of Medicine, National Institutes of Health, https://
    profiles.nlm.nih.gov/ps/retrieve/Narrative/CD/p-nid/260 (accessed
    Oct. 11, 2016).

17. Paul Berg, David Baltimore, et al., "Summary Statement of the

Asilomar Conference on Recombinant DNA Molecules," *Proceedings of the National Academy of Sciences of the United States of America* 72, no. 6 (June 1975): 1981–84.

18. Interview with Paul Berg, June 6, 2016.

19. David Baltimore, Paul Berg, et al., "A Prudent Path Forward for Genomic Engineering and Germline Gene Modification," *Science* 348, no. 6230 (Apr. 3, 2015): 36–38.

20. David Cyranoski and Sara Reardon, "Chinese Scientists Genetically Modify Human Embryos," *Nature*, Apr. 22, 2015, doi:10.1038/nature.2015.17378.

21. Sarah Zhang, "CRISPR Is Getting Better. Now It's Time to Ask the Hard Ethical Questions," *Wired*, Dec. 1, 2015, www.wired.com/2015/12/stop-dancing-around-real-ethical-problem-crispr (accessed Oct. 11, 2016).

22. Michael Specter, "The Gene Hackers," *New Yorker*, Nov. 16, 2015, www.newyorker.com/magazine/2015/11/16/the-gene-hackers (accessed Oct. 11, 2016.)

23. Gregory Stock, *Redesigning Humans: Choosing Our Genes, Changing Our Future* (New York: Houghton Mifflin, 2003).

24. Stephen D. H. Hsu, "On the Genetic Architecture of Intelligence and Other Quantitative Traits," *arXiv*, Aug. 30, 2014, arXiv:1408.3421v2 (accessed Oct. 11, 2016).

25. Interview with Jamie Metzl, Apr. 28, 2016.

26. Steven Olson, ed., *International Summit on Human Gene Editing: A Global Discussion* U.S. (Washington, DC: National Academies Press, 2016), doi: 10.17226/21913.

27. David Baltimore, Francoise Baylis, et al., "On Human Gene Editing: International Summit Statement," press release, U.S. National Academies of Science, Engineering, and Medicine, Dec. 3, 2015, www8.nationalacademies.org/onpinews/newsitem.aspx?RecordID=12032015a (accessed Oct. 11, 2016).

INDEX